Short Introductions to Cultural Heritage Science

Series Editor
Peter Vandenabeele
Department of Archaeology
Ghent University
Ghent, Belgium

This book series addresses people in humanities (mainly archaeologists, art-historians, conservation scientists, anthropologists, historians, etc.) who are interested in the technical aspects of materials and/or objects.

As this series aims to fill the gap in knowledge on technical aspects of different materials, readers should have a better understanding after studying the 'introductions' included in the series, which will refer to more advanced texts on the topic at hand.

The series is didactic and provides sufficient background information to understand the different aspects of the materials studied. This gives readers the opportunity to reach a level that allows them to interact with specialists or to understand scientific papers in the specific research domain.

More information about this series at http://www.springer.com/series/15439

Stefanos Karampelas • Lore Kiefert
Danilo Bersani • Peter Vandenabeele

Gems and Gemmology

An Introduction for Archaeologists, Art-Historians and Conservators

Stefanos Karampelas
Laboratoire Français de
Gemmologie (LFG)
Paris, France

Danilo Bersani
Department of Mathematical,
Physical and Computer Sciences
University of Parma
Parma, Italy

Lore Kiefert
Gubelin Gem Lab
Lucerne, Switzerland

Peter Vandenabeele
Department of Archaeology
Ghent University
Ghent, Belgium

Short Introductions to Cultural Heritage Science
ISBN 978-3-030-35451-0   ISBN 978-3-030-35449-7   (eBook)
https://doi.org/10.1007/978-3-030-35449-7

© Springer Nature Switzerland AG 2020, corrected publication 2021
This work is subject to copyright. All rights are reserved by the Publisher, whether the whole or part of the material is concerned, specifically the rights of translation, reprinting, reuse of illustrations, recitation, broadcasting, reproduction on microfilms or in any other physical way, and transmission or information storage and retrieval, electronic adaptation, computer software, or by similar or dissimilar methodology now known or hereafter developed.
The use of general descriptive names, registered names, trademarks, service marks, etc. in this publication does not imply, even in the absence of a specific statement, that such names are exempt from the relevant protective laws and regulations and therefore free for general use.
The publisher, the authors, and the editors are safe to assume that the advice and information in this book are believed to be true and accurate at the date of publication. Neither the publisher nor the authors or the editors give a warranty, expressed or implied, with respect to the material contained herein or for any errors or omissions that may have been made. The publisher remains neutral with regard to jurisdictional claims in published maps and institutional affiliations.

This Springer imprint is published by the registered company Springer Nature Switzerland AG.
The registered company address is: Gewerbestrasse 11, 6330 Cham, Switzerland

Acknowledgments

The authors wish to thank the Swiss National Museum and specifically Dr. Tiziana Lombardo and Fabian Muller (Affoltern am Albis and Zurich, Switzerland), Markus Bamert (Einsiedeln Abbey, Einsiedeln, Switzerland), Dr. Hugo Miguel Crespo (University of Lisbon, Portugal), Dr. Rui Galopim de Carvalho (Portugal Gemas Academy, Lisbon, Portugal), Enzo Liverino (Torre del Greco, Italy), and the Historical Museum Basel (Switzerland) for providing the images of some items. SK would like to thank his co-workers for the photos they took at LFG, as well as Pr. Emmanuel Fritsch (IMN-CNRS, Nantes, France), Pr. Benjamin Rondeau (University of Nantes, France), and Dr. Jaroslav Hyrsl (Prague, Czech Republic) for providing useful and difficult to find references. LK would like to thank her co-workers for the photos they took at Gübelin Gem Lab.

Contents

1 Introduction 1
2 Gems Through the Ages 5
3 Gem Analysis 39
4 Gem Treatments, Synthetics and Imitations 67
5 Archaeometrical Questions (Case Studies) 91
Correction to: Gems and Gemmology C1
Glossary 105
Index 111

About the Authors

Stefanos Karampelas is Chief Gemmologist at the Laboratoire Français de Gemmologie (LFG), Paris, and he is also lecturing for the Advanced Gemmology Diploma at the University of Nantes, France. He started studying Geology and Mineralogy at Aristotle University of Thessaloniki (Greece), he has MSc in Geosciences and Advanced Gemmology Diploma, both from the University of Nantes (France), and he completed his PhD in Materials Physics at the University of Nantes (France) and Mineralogy at the Aristotle University of Thessaloniki (Greece) on the non-destructive study of the origin of pearls' color. He worked as a Research Scientist for about 7 years at Gubelin Gem Lab and a further year at GemResearch Swisslab, both in Switzerland, as well as for about 3 years as Research Director for the Bahrain Institute for Pearls and Gemstones (DANAT). His research interests include advanced non-destructive techniques applied to all kind of gem materials. He has published numerous peer-reviewed articles in scientific journals, contributed to books, and visited several gem mines as well as natural and cultured pearl producing areas around the globe. He is also frequently delivering lectures to international scientific conferences and gemmological meetings. He is a Member of the Commission of Gem Materials of International Mineralogical Association and of the Editorial Board of *Gems & Gemology*, Delegate for the International Gemmological Conference, and Associate Editor of *The Journal of Gemmology*.

Lore Kiefert is Chief Gemmologist at Gübelin Gem Lab, where she is, among other tasks, responsible for the training of gemmologists. She started studying mineralogy in Heidelberg, Germany, in 1981 and completed her master's thesis on the origin of sapphires in 1987. She then moved to Australia to study the mineralogical and chemical composition of desert dust, which earned her a PhD in 1996. Leaving Australia for Switzerland in 1994, she joined the SSEF Swiss Gemmological Institute as Deputy Director and went on to become its Director of the Coloured Stones Department as well. During her time at the SSEF, she completed her FGA Diploma in 1998. In 2005, she moved to New York to head the AGTA Gemmological Testing Center as Laboratory Director until she decided to return to Europe to join the Gubelin Gem Lab as Chief Gemmologist in October 2009.

She has authored and coauthored over 100 publications in gemmological and scientific journals, as well as chapters in textbooks such as the Handbook of Raman Spectroscopy. She regularly delivers gemmological lectures at conferences worldwide and has co-organised two gemmological conferences in Switzerland and the USA. She is also on the editorial review board of *Gems & Gemology* and *The Journal of Gemmology*, as well as Member of the LMHC. She was awarded Professorship at Tongji University in Shanghai, China, in 2017, where she regularly conducts workshops and lectures.

Danilo Bersani is Associate Professor in Physics at the University of Parma, Italy, Department of Mathematical, Physical and Computer Sciences. His research is mostly devoted to the spectroscopic analysis of gems, minerals, nanocrystalline materials, objects related to art and archaeology, in particular by means of Raman spectroscopy. He is Author of more than 150 scientific publications in international journals and has given more than 250 presentations at international conferences. He is also Organiser and Member of the scientific committees of different international conferences, such as GeoRaman, inArt, and RAA (Applications of Raman Spectroscopy in Art and Archaeology).

Peter Vandenabeele is Professor in Archaeometry at Ghent University, Belgium. He is Member of the Department of Archaeology and Associated Member of the Department of Chemistry. In 2000, he obtained his PhD in Analytical Chemistry with research on the application of Raman spectroscopy and total reflection X-ray fluorescence for the analysis of art objects. His research mainly focusses on the development and optimisation of spectroscopic techniques for archaeometrical applications. He authored the handbook *Practical Raman Spectroscopy: An Introduction* and coedited several other books. He is Author of peer-reviewed research papers and frequently presents his recent research in international scientific conferences. Moreover, he is Chairman and Committee Member of international conferences in the fields of art analysis and Raman spectroscopy.

The original version of this book was revised: The biography of Stefanos Karampelas has been updated, as well as his affiliation. Further, the Acknowledgements section now includes the names of those who took the photos for the replacement figures. The correction to this book is available at https://doi.org/10.1007/978-3-030-35449-7_6

Chapter 1
Introduction

In Archaeology and Art History, researchers are often confronted with gems of all types and ages. As gems are not man-made and most of them have formed thousands to millions of years ago, they have distinctly different properties, but also distinctly different problems, than monuments or artwork formed by humans. An archaeologist's concern about the latter besides verification of its authenticity is its preservation. In gemmology, preservation is not a major issue except for some organic gems such as pearls or amber. However, certain properties of gems, such as their geographic origin, cutting style and methods of treatment, can give valuable information about the authenticity and provenance of the jewellery piece.

People involved in the analysis of gems, starting from gem amateurs to professional gemmologists, including archaeologists, art historians, conservators, mineralogists, and gem dealers have a very complex task. They should not only understand the basics of many disciplines (mineralogy, crystallography, geology, chemistry, physics, and sometimes biology) but also take into consideration the economic aspects. The analysis of gems starts from the identification of their chemical composition, determining whether they are natural or "artificial" (*i.e.*, imitation or synthetic), checking for enhancement treatments, grading and sometimes determining their geographic origin. In addition, all the information should be obtained by applying non-destructive and non-invasive methods. Analysis of such gems becomes more complex when they are mounted or embedded in jewels or artworks. If the gems and jewels to be tested are preserved in collections and museums, it is often impossible to remove from their setting due to their high value and to avoid any damage, therefore such gems will require the use of *in situ* techniques for their identification.

Gemmology is the science dealing with gems. Therefore, it is extremely important for the gemmologist handling the gems to be professionally trained on the various methods of identification taking into account all the above-mentioned situations and tasks. In order to study gemmology, several gemmological laboratories offer private courses of various durations and qualities. Academic degrees in gemmology are scarce; and only few geological/mineralogical university departments around the globe offer gemmological courses in their degree program. Gemmology is, in general, considered a geoscience and a branch of mineralogy and its roots can be traced back to Theophrastus (315 BCE) who described how minerals and gems grow and to Pliny (79 CE) who mentioned identification issues of gems. At the beginning of the nineteenth century, R-J Haüy and his contemporaries started to develop gemmology as a modern science. When the first synthetic gems entered the market at the turn of the nineteenth to twentieth century, gemmology further developed as a separate science. During the same period, special gemmological microscopes were developed. Other instruments such as the polariscope, dichroscope and refractometer, which are still the standard instrumentations for gemmology, were already developed in the course of the nineteenth century (Ferguson and Brewster 1823; Abbe 1874). At the beginning of the twentieth century, the first gemmological laboratories appeared in the United Kingdom, France, the United States of America and Germany, soon followed by a private laboratory in Switzerland. Most of these laboratories also offered educational programmes. Parallel to the appearance of these laboratories, more gemstone mines were discovered worldwide, more synthetics were produced, and various types of treatments were applied on gems. By the 1970's, it was essential for important pieces of jewellery with gems to be accompanied with a report stating the nature of the gem as well as its treatment status. Prior to this development, gem reports stating the quality of diamonds were already issued in the 1950's. Nowadays, a multitude of gemmological laboratories exist, and nearly every gem needs to be tested. The major gemmological laboratories are now equipped with more and more advanced instruments to assist in identification and to meet the challenges of the modern gem market, where treatments are ubiquitous, and origin of a gem became more important than in the past.

In parallel, analytical techniques for archaeological and art-historical investigations have greatly improved over the past two decades, and often the same instruments can be used for several applications including gems. In addition, chapters in interdisciplinary handbooks describing some of these techniques, have become more popular. These handbooks, as well as international archaeological and art-historical conferences, contain chapters and sessions about gems (Kiefert et al. 2001, 2005, 2012, 2019; Karampelas and Kiefert 2012; Fritsch et al. 2012). However, none of these are solely dedicated to the challenges that archaeologists and art historians face when confronted with gems and their analysis. The current book is an attempt to consolidate knowledge about the history of gems, including synthetic gems and imitations, as well as of their treatments, together with analytical techniques in a form that makes it possible for archaeologists and art historians to draw on this gemmological knowledge.

People starting to work on the analysis of gems should be aware of the multitude of questions posed by the characterization of a gem. This book projects details of those aspects and instrumentation that can be used, at different levels, to obtain the desired answers. It starts with a chapter on the history of gems, reaching as far back as Palaeolithic and Neolithic and describing the appearance of all major gemstones, organic gems and other gem materials over time. It also covers their changing nomenclature as well as the cutting styles over the centuries. Knowledge about these factors assists in establishing a time frame to when the gem was probably used. This is followed by an extensive chapter describing not only the classical gemmological methods but also the advanced instrumentation. Details about microscopy, and the use of dichroscope, polariscope, hand spectroscope and refractometer are described in great detail. The principles of UV-Vis-NIR spectrometers, Raman spectrometers, FTIR spectrometers and XRF instruments as well as X-ray imaging, LIBS and Laser Ablation ICP-MS and other instruments such as the Diamond View are explained, followed by instruments that are usually only available at Universities such as PIXE or SIMS. Chapter 4 describes gem treatments and their developments over time, starting with the oldest methods like dyeing and foiling, and evolving in the more recent treatments which include heat treatment, glass filling and chemical element diffusion. This is followed by an insight in the development of imitations as well as synthetics. The chapter attempts also to list them chronologically, to enable archaeologists and art historians to date their observations. For example, a heat-treated sapphire is unlikely to be found in a piece of jewellery from Roman times.

The last chapter of the book shows some examples of gem testing and case studies of gems in the fifteenth century from the Basel treasure, ecclesiastical objects from the sixteenth century from the Monastery of Einsiedeln, two historic blue diamonds and a sapphire from the French Crown Jewels of Louis XV. These objects are mainly in museums and special care must be taken when analysing them, sometimes using hand-held instruments that can be applied in-situ. The chapter goes further to give examples of isotopic analyses on a series of ancient emeralds and natural pearls. Oxygen isotopic data provides useful information on origin determination of emeralds and carbon isotopes are used for age determination of natural pearls, and conclusions on the provenance can be drawn.

References

Abbe E (1874) Neue Apparate zur Bestimmung des Brechungs- und Zerstreuungsvermögens fester und flüssiger Körper. Jenaische Z Naturwiss, 8 NF I:96–174
Ferguson J, Brewster D (1823) Lectures on select subjects in mechanics, hydrostatics, hydraulics, pneumatics, optics, geography, astronomy and dialling, vol II, 3rd edn. Stirling & Slade, and Bell & Bradfute, Edinburgh
Fritsch E, Rondeau B, Hainschwang T, Karampelas S (2012) Raman spectroscopy applied to gemmology. In: Dubessy J, Caumon M-C, Rull F (eds) Applications of Raman spectroscopy to earth sciences and cultural heritage, vol 12. European Mineralogical Union and Mineralogical Society of Great Britain & Ireland, EMU Notes in Mineralogy, pp 453–488
Karampelas S, Kiefert L (2012) Gemstones and minerals. In: Edwards HGM, Vandenabeele P (eds) Analytical archaeometry: selected topics. Royal Society of Chemistry Publishing, Cambridge, pp 291–317
Kiefert L, Hänni HA, Ostertag T (2001) Raman spectroscopic applications to gemmology. In: Lewis IR, Edwards HGM (eds) Handbook of Raman spectroscopy. Marcel Dekker, Inc, New York, pp 469–489

Kiefert L, Chalain JP, Häberli S (2005) Case study: diamonds, gemstones and pearls: from the past to the present. In: Edwards HGM, Chalmers JM (eds) Raman spectroscopy in archaeology and art history, vol XXI. Royal Society of Chemistry, Cambridge, pp 379–402

Kiefert L, Epelboym M, Kan-Nyunt HP, Paralusz S (2012) Applications to the study of gems and jewellery. In: Chalmers JM, Edwards HGM, Hargreaves MD (eds) Infrared and Raman spectroscopy in forensic science. Wiley, Chichester, pp 455–468

Kiefert L, Hardy P, Schollenbruch K, Xu W (2019) New case studies: diamonds, jades, corundum and spinel. In: Vandenabeele P, Edwards H (eds) Raman spectroscopy in archaeology and art history. Royal Society of Chemistry, Cambridge, pp 254–270

Chapter 2
Gems Through the Ages

Since thousands of years, gems were used as tools and amulets often associated with social status and money. The first use of gems goes back to the age of primitive human beings. During different periods in time, different cultures were considering different materials as gems; for example: animal bones, mollusc shells as well as tinted glass, rock pieces (e.g., soapstone, chlorite, lapis lazuli, nephrite 'jade'), different forms of silica (quartz, amethysts, agate, chalcedony, opal etc.), amber and pearls.

The first gems which were used as found were mostly translucent to opaque (rarely transparent) and in some instances they were drilled to be strung with the use of thread or silk. In the early days of the western civilization some gems were also engraved. Transparent stones became more popular when humankind first started understanding the various techniques of cutting and polishing facets on the different kinds of gems found. Thus, gems widely used today such as ruby, blue sapphire and diamond became popular at a later stage. Some the famous mines where highly prized gems are mined, such as Colombia for emeralds and Kashmir (Sumjam, Kundi Valley) for blue sapphires were found relatively recently, after the fifteenth century and at the end of the nineteenth century respectively. Some popular gems such as tanzanite (violet-blue to blue gem quality zoisite) and "Paraiba-type" tourmaline ("neon" blue to green Cu-and Mn- bearing Na- or Ca-rich tourmaline) have very limited archaeological importance as they were found after the half of the last century; late 1960s in Tanzania and late 1980s in Brazil respectively.

The first person to learn how to polish rough surfaces of gems remains unknown. It is well possible that the earliest attempts may have been to give a sort of polish to the irregular surfaces by rubbing one stone against another in running water (Webster and Anderson 1983). The earliest forms of cutting gave a curved surface (a style known today as cabochon) or simply a flat surface upon which the worker could engrave. At the beginning, gems were used in their rough form, or polished with curved surfaces or engraved (Figs. 2.1 and 2.2). Point cut diamonds which involved a simple polishing following their natural faces were also known in India before the

---

The original version of this chapter was revised: New figures are updated with new captions and few figures are updated without changing their captions. The correction to this chapter is available at https://doi.org/10.1007/978-3-030-35449-7_6

Fig. 2.1 Two gold rings with intaglios of about 2 cm (longest dimension) made of cryptocrystalline silica gems (left photo onyx and right photo black onyx) from Roman period. (Photos: Swiss National Museum)

Fig. 2.2 A cameo about 3 cm (longest dimension) made of cryptocrystalline silica gem (not tested by the authors) from Roman period and later engraved in Renaissance Florence with 'LAV.R.MED', while in the collection of Lorenzo de' Medici, the Magnificent (1449–1492). Museo Archeologico Nationale, Naples, Italy, inv. 25,851 (Arditi 124). (Photo by Hugo Miguel Crespo)

tenth century CE. Gem cutting started at the start of thirteenth century (e.g., table cut in diamonds was obtained by grinding away the tip of an octahedral crystal, which produced a flat table facet; Figs. 2.3 and 2.4) and it radically improved after the industrial revolution (Klein 2005). Nowadays, the cutting styles significantly help in revealing the gems' appearance, for example a perfect round brilliant cut diamond shows better dispersion (fire); one of the first diamond cut with faces is the

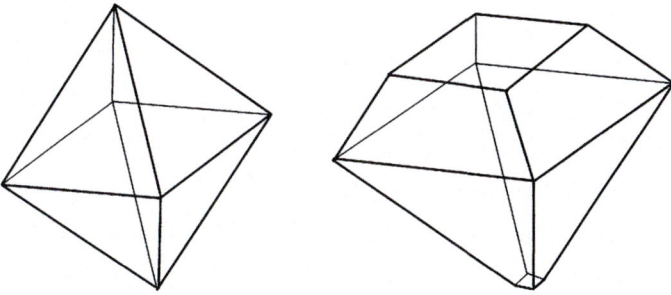

Fig. 2.3 Early cuts of diamonds; point cut at the left and table cut on the right

Fig. 2.4 A gold ring adorned with six smaller and one larger (about 6 mm in longest dimension) table cut diamonds, presumably from India (not tested by the authors); seventeenth century, Ashmolean Museum, Oxford, England, inv. WA1899.CDEF.F818. (Photo by Hugo Miguel Crespo)

rose cut (Fig. 2.5). Several gemmological terms are used to describe the cut of a gem; for shape (round, square etc.) and cut style (step, brilliant etc.; Fig. 2.6).

Gem engraving is a miniaturist art form whereby designs are either cut into or on the surface of a gem (cameo and intaglio respectively). An intaglio is made by grinding away material below the surface of the gem, leaving an inverse image. A cameo is the opposite of an intaglio: that is, the subject is sculpted above the surface of the gem, appearing in relief on stones usually of two or more different-coloured layers. Another type of engraved gem occasionally encountered is the chevet, in which a raised figure rests in a background sunk below the surface. However, this is a relatively modern development (Gray 1983). The outstanding feature of gem carving is the small scale of the art, as it is done most often on a surface less than 2 cm in diameter giving great attention to the detail accomplished on such a small surface. The use of metal points with emery or another hard powder (added as the drill revolves to get the actual abrading) was introduced at around 2000 BCE and thus most gems were easier to carve. Engraving exists for more than 8000 years; it is

Fig. 2.5 A highly decorated gold stomacher (15.5 cm at its longest dimension) with numerous rose cut diamonds (not tested by the authors). Portugal, early eighteenth century. (Photo by Carlos Pombo Monteiro; © Fundação Eugénio de Almeida/Archidioceses of Évora, Portugal)

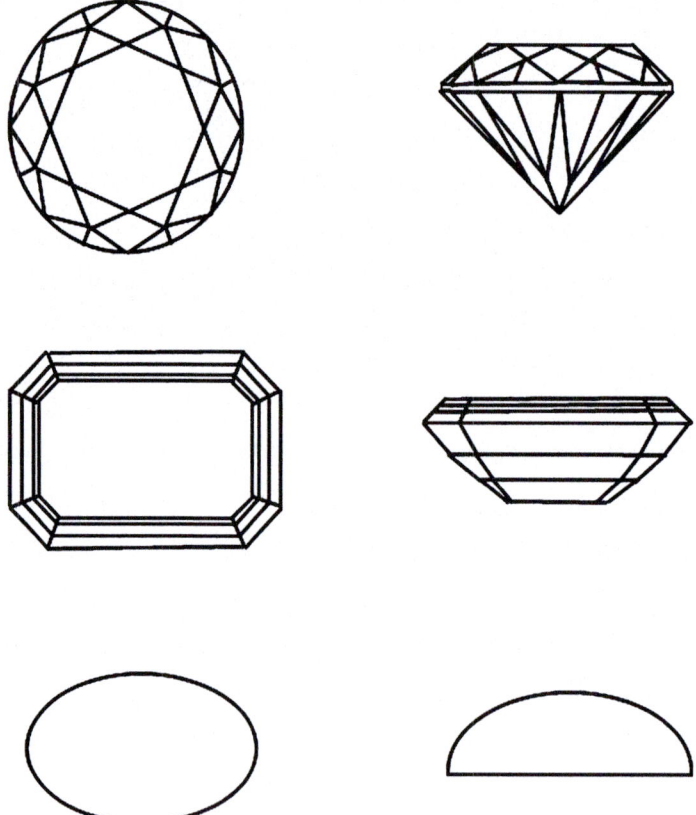

Fig. 2.6 Different gem cut styles, on the left the top view and one right the side view of brilliant cut (top), step cut (middle) and cabochon (bottom)

believed that the first seals of various shape and from various materials were probably introduced between 6500 and 6000 BCE in the Neolithic cities of Mesopotamia (Gray 1983). In addition, some people believed that the art of gem carving dates at least back 7000 BCE in the Hindus valley (Rapp 2002a).

Cylinder seals were the most common, however the Egyptians also produced scarab seals. These seals took their convex shape from the scarab beetle (with a religious significance) and on the flat bottom side the artisan engraved hieroglyphic characters which served as the bearer's mark. From approximately 3200 to 200 BCE, Egypt was considered to be the world's main producer of engraved gems; for example, carnelian and other forms of cryptocrystalline silica, rock crystal and amethyst were used more often than turquoise, emerald and peridot (Gray 1983). Minoan and Mycenaean cultures worked mostly on hard material such as quartz and chalcedony before 1100 BCE, some of lenticular shape (a.k.a. island gems; Middleton 1891). It seemed that the ancient Greeks learnt later from the Phoenicians the engraving and scaraboid appearance (scarab back was replaced by a simple unornamented dome). Amethyst and garnet became particular favourites and were often used for fine pieces, and interestingly glass was also highly valued (Gray 1983). During the Hellenistic and Roman years, the scaraboid evolved in the ringstone; where the seal was mounted and shown at all times and became ideal for personal signets. The acquisition of tools such as the diamond point in a rod from India, generally facilitated the engraving process. In parallel, the engraving of cameos sometimes done on a gem with different colour layers, was improved (see again Fig. 2.1). By the conquest of the Eastern territories by Alexander the Great and the Romans, new material was available for carving which included garnets (almandine, hessonite and rhodolite), amethyst, rock crystal, as well as corundum and beryl (Gray 1983). Middle Ages showed little contribution to gem engraving and the Renaissance period was the time when faceted gems started to appear.

A problem that researchers frequently face whilst studying gems of archaeological importance is the nomenclature used in the old text and the current gemmological literature. A good example of this is peridot, used by gemmologists for the gem quality yellowish-green to green olivine (($Mg,Fe)_2SiO_4$). In old texts this gem was called topazios and was also known as chrysolite (Gubelin 1981). Topazios was the primary name given to the island of Zabargad in the Red Sea (a.k.a. St John's Island), which today belongs to Egypt and was a source of peridot at those times (and called topaz). At the early eighteenth century, the name topaz was affixed to a mineral with the chemical formula $Al_2SiO_4(F,OH)_2$ and the name chrysolite was no longer used because it was noticed that it was the same mineral as olivine. Another good example is the nomenclature used for silica varieties (for more information see the paragraph below on "silica gems"). The exact geographic origin of a gemstone was also an issue, as in some manuscripts there was often a confusion between the place where the gem was purchased and the place where it was mined. This chapter focuses mainly on important gems which are used nowadays. Materials that are not used as gems today (e.g., soapstone, chlorite, hematite) will not be covered.

2.1 Pearls

Natural pearls are probably the oldest gems known to humankind, perhaps because no shaping or cutting is required (Webster and Anderson 1983; Dirlam et al. 1985). Pearls are organogenic (biogenic) gems composed of calcium carbonate ($CaCO_3$), organic matter (β-chitin and an assemblage of glycoproteins, formerly known as conchiolin) and water, formed by a mollusc with a shell. Seafood used to be the primary food for people living on the coasts and islands all over the world; this was how pearls were found. Virtually all molluscs with a shell, living in saltwater and freshwater (rivers and lakes) can produce pearls. The vast majority of pearls are found in bivalve molluscs (Class: Bivalvia, Linnaeus, 1758); *i.e.*, their bodies are enclosed by a shell consisting of two hinged parts. The bivalves in which most of pearls are found belong to *Pinctada* (Röding, 1798) genus and are called "pearl oysters". Natural pearls are also sometimes found in univalves such as gastropods (Class: Gastropoda, Cuvier, 1795) and in extremely rare cases in cephalopods (Class: Cephalopoda, Cuvier, 1797). Most of pearls are nacreous, *i.e.*, composed by alternating concentric layers of aragonite (sometimes with small regions made of calcite and/or vaterite), organic matter and water and rarely non-nacreous, *i.e.*, all pearls present other than nacreous structure (*e.g.*, composed by fibres of aragonite and/or calcite).

Archaeological excavations confirm the use of pearls in regions around the Californian and Arabian Gulf more than 8500 and 7500 years ago respectively; with the island of Bahrain (also known as the land of Dilmun in antiquity) playing an important role in pearl trading. The oldest pearls (sized <4 mm) found till today was dated back to *ca.* 6500 BCE (by doing radiocarbon dating on pearls) and found in Espiritu Santo Island, Baja California Sur, Mexico (Fujita et al. 2017; Ainis et al. 2019). In, nowadays, UAE the oldest finding was a pearl (< 4 mm) dated back to *ca.* 5500 BCE (age estimated via graveyard's dating; Charpentier et al. 2012). Other pearls, also nacreous, of different Neolithic ages (5200–3500 BCE) were also found during excavations in South-Eastern Arabia. The oldest Mesopotamian example comes from Uruk and dates back to *ca.* 3200–3000 BCE (Donkin 1998).

Surprisingly and contrary to what is believed and sometimes written, pearls did not have universal admiration. Only few pre-Columbian jewels with pearls were found; it is most likely that pearls started to become popular only around 400 BCE in the western regions and references on these are found in Theophrastus (Kunz and Stevenson 1908; Donkin 1998). Conquests of Alexander the Great and the rise of the Ptolemaic empire played an important role in the natural pearl trade. It was believed that other than saltwater pearls from the Arabian Gulf and Red Sea, some freshwater pearls derived from freshwater molluscs from rivers in Britain were traded (Strack 2008).

It is around the same period of time that the first references for pearls in Arthashastra where several sites of pearl fishing in the far South of India and on the North West coast of Sri Lanka were reported (Bari and Lam 2009). In ancient Egypt, pearls became more important during and after the time of Ptolemy I. It is also believed that in China and Japan pearls existed from ancient times. In the book of

Shu King, it was mentioned that Yu the Great (earlier than 2000 BCE) received a natural freshwater pearl fished from a mollusc found in Hwai or Huai river and one necklace made out of various shaped natural pearls from the province of Zhejiang. By 200 BCE, traders had established the Silk Road to and from China. The demand of Chinese silk, jade and other products was mirrored by the demand within China for natural pearls. Pearl fisheries (from *P. fucata*; closely related to *P. radiata*) have existed since 700 CE in Japan in the district in Omura Bay near Nagasaki; however most natural pearl fishing took place after the seventeenth century (Bari and Lam 2009). Some researchers also believe, that the discovery of ancient pearls during excavation required fine sieving of sediments, which is not always done and that's is probably a reason why they were rarely found (Charpentier et al. 2012).

The earliest archaeological evidence of pearls in jewellery was found at Susa, the ancient capital of Elam, in the Khuzistan region of Iran (Kunz and Stevenson 1908). The necklace is currently in the Louvre museum in Paris. It is made up of three equal rows consisting of a total of 216 pearls of various shapes (<5 mm diameter). It was recovered from the bronze sarcophagus of an Achaemenid princess at Susa and dates to a period not later than the fourth century BCE. Another important early archaeological piece is the Paphos pin, currently displayed in the British Museum in London, dating to the third century BCE (adorned with a saltwater pearl of 14 mm and a freshwater pearl of 4 mm; Dirlam et al. 1985). A natural 2000 years old saltwater pearl of 5.9 mm diameter (probably from *P. albina*), dated via the other excavation findings, was located in Kimberley in Australia (Szabo et al. 2015).

Pearls became popular and abundant after annexation of Syria by Pompey the Great and also have been cited as one of the reasons for the first Roman invasion of Great Britain in 55 BCE (Dirlam et al. 1985). Mid-third century, byzantine jewellers started to use fine gold leaf with the most popular gems at the time, like saltwater pearls from the Arabian Sea, agate, pink quartz, amethyst, lapis lazuli from central Asia and emerald principally from Egypt. The mosaic picture of Ravenna's St Vital Cathedral showed Theodora (wife of Byzantine emperor Justinian I) wearing pearls.

Natural freshwater pearls found in Europe in *Margaritifera margaritifera* (this mollusc is in the IUCN red list of threatened species) have been known since Roman times (Strack 2006). These pearls are relatively small, of white to pinkish colour, and were found in the rivers throughout central and northern Europe with the more important ones located in Scotland, Russia, Germany and France. Some ecclesiastic items as well as some folkloric customs were adorned with these pearls. In some cases, fresh and salt-water pearls were mixed and in the year 1355 jewellers were forbidden to use saltwater ('oriental') pearls and freshwater ('river') pearls in the same piece. "Seed" natural pearls (i.e., < 3 mm diameter) were also widely used. In China, during the Ming Dynasty (1368–1644) natural pearls were highly appreciated (Strack 2008).

Columbus' travels in America played an important role, among others, for gems and also pearls. Natural saltwater pearls from *P. imbricata* (a bivalve closely related to *P. radiata*) or from another *Pinctada* sp. of white colour from Nueva Sparta state located off the coast of present-day Venezuela as well as dark coloured pearls from

Fig. 2.7 From left to right, cuts of a cultured saltwater pearl from *Pinctada maxima* with bead (about 12.1 mm in length; the white coloured bead, made out from freshwater shell, with a diameter of 6.5 mm), cultured saltwater pearl with bead from *Pinctada fucata* (diameter of about 5.5 mm; the white coloured bead, made out from freshwater shell, with a diameter of 5 mm), cultured freshwater pearl without bead from *Hyriopsis cumingi* and a natural saltwater pearl from *Pinctada radiata*. (Photo: Bérengère Meslin Sainte Beuve/LFG)

Pinctada mazatlanica and *Pteria sterna* were fished off the Mexican and Panama waters on the Pacific side and were brought back to Europe (Carino and Monteforte 1995). During this period, saltwater pearls coming from the Middle East were known as 'oriental' pearls and those coming from Central America were known as 'occidental' pearls. Dark coloured pearls from Polynesia, found mainly in the *Pinctada margaritifera*, were fished before the European explorers arrived in midsixteenth century (Goebel and Dirlam 1989). However, little was known about how the pearls were used by the native Polynesians or the early European visitors. Jewellery pieces with such pearls were not dated before mid-nineteenth century. Jean-Baptiste Tavernier also brought natural pearls from his seventeenth century voyages.

In Australia and Oceania, systematic fishing of natural pearls, mainly from *P. maxima*, started slightly after the European settlement in the early nineteenth century (Scarratt et al. 2012). The peak of the natural pearl fishing was around the middle of nineteenth century to the start of twentieth century and occasionally resulted in the depletion of the molluscs. The beginning of the twentieth century started the big scale pearl cultivation. Cultured pearls are formed through human intervention by graft transplantation, without (cultured pearl without bead) or with (cultured pearl with bead) simultaneous solid nucleus (usually cut out from a freshwater mollusc's shell) implantation, nowadays on molluscs raised in farms (wild molluscs at the old times); see some examples in Fig. 2.7. In today's world market 99.8% of the pearls are cultured with 95% being freshwater cultured pearls without bead (most of them are cultivated in Chinese rivers and lakes using *Hyriopsis cumingi* bivalve).

However, cultivation of pearls was known for a much longer time. Cultured blister freshwater pearls were produced since the thirteenth century in China and round freshwater cultured pearls were produced since the eighteenth century in China and Sweden (Scarratt et al. 2000). Until then, no big scale cultivation operations were known. All the pearls described above have a nacreous structure appearance. Non-nacreous natural pearls were also found in jewellery at the beginning of the nine-

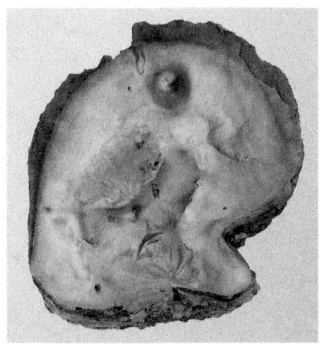

Fig. 2.8 Natural blister attached in the inner part of a *Pinctada margaritifera* shell. Width of the shell is 10 cm. (Photo: Stefanos Karampelas/LFG)

teenth century with 'queen conch' pearls (pink pearls found in *Lobatus gigas* gastropod in the Caribbean Sea; Bari 2007). However, three to four centuries older cameos made of gastropod shells were also found. Both nacreous and non-nacreous pearls can be blister; *i.e.*, they have been naturally attached to the inner wall of the shell. The subsequently formed layers of nacreous or non-nacreous material are continuous with those of the inner wall of the shell. On the other hand, natural blisters are not pearls but are merely domed formations on the inner part of the shell (Fig. 2.8).

2.2 Gem Corundum (Rubies and Sapphires)

Rubies and sapphires are coloured gem varieties of the mineral corundum (α-Al_2O_3). Ruby is the red variety, sapphire is referred to the blue colour varieties, while all other coloured sapphires are called fancy and need the colour prefix (e.g., pink sapphire, yellow sapphire, etc.), with padparadscha sapphire (pinkish-orange to orangey-pink sapphire) as the only exception. It is believed that in Theophrastus the term anthrax was related to ruby (also attributed to spinel and red garnet) and in Pliny all red stones were called carbunculus (Giuliani et al. 2014). Commonly, rubies were mixed with red garnets (pyralspite series) and spinel (which was properly identified at the end of the eighteenth century). Up to the thirteenth century, variations of the term sapphire referred to all blue stones, especially lapis lazuli. The term hyacinth was believed by some people to describe blue sapphires and was used by others to describe the fancy coloured ones (e.g., yellow). The oldest historical corundum gems originated from alluvial deposits in Sri Lanka (Ceylon); which is still considered to be one of the main producers (Fig. 2.9). The stones found in that locality were mostly blue and rarely yellow, pink (very rarely red towards ruby), purple, with colour change as well as some star sapphires. The first corundum travelled probably to

Fig. 2.9 Rough corundum crystals from Sri Lanka; right crystal is 1.4 cm in length. (Photo: Stefanos Karampelas/LFG)

Europe during the period of Alexander the Great but sapphires were mostly used during Roman times. A majority of them were used as beads (sometimes polished; Fig. 2.10) and often drilled along the longest axis, however, some stones were also engraved (Vollenweider 1974). One of the oldest cuts (ninth century CE) was the pyramidal shape cut sapphire; a good example can be found on the Berengar cross. During the thirteenth–fourteenth century sapphires started to be faceted in different shapes (Fig. 2.11). Between the tenth and fifteenth centuries, with another short mining period in the nineteenth century, small (usually below 1 cm) blue sapphires were used in jewellery in France (Le Puy; Hyrsl 2001b). Sapphires from Jizerska Louka (Czech Republic, formerly Bohemia) were also believed to have been exploited about the same time, without having precise information (Mocquet 2003).

From the sixth century onwards, rubies from Burma, currently Myanmar, were used, however, some researchers stated that these only entered the market after the twelfth century. Burmese sapphires were known ever since the thirteenth century (Hughes and Thoresen 2017). Afghani ruby and sapphires from the mines of Jegdalek have been worked since the thirteenth century (Bowersox et al. 2000). Additionally, corundum from Siam (now Thailand and Cambodia) was used since the fifteenth

2.2 Gem Corundum (Rubies and Sapphires)

Fig. 2.10 Two gold rings with a sapphire (upper photo; about 12 mm in longest dimension) and a ruby (bottom photo; about 8 mm in longest dimension), both cabochon, presumably from Sri Lanka and Burma respectively (not tested by the authors). Upper photo: thirteenth–fourteenth century, Venice, Italy. Ashmolean Museum, Oxford, England, inv. WA1899.CDEF.F804. Bottom photo: Late sixteenth century, Ashmolean Museum, Oxford, England inv. WA1899.CDEF.F475. (Photos by Hugo Miguel Crespo)

century, with some people believing that this goes back to as far as the early thirteenth century (Hughes and Thoresen 2017). Blue sapphires from Kashmir mountains were also mined at the end of the nineteenth century (Atkinson and Kothavala 1981). In the twentieth century many new gem corundum deposits were found with the most important ones been in Madagascar and Mozambique (Africa) as well as some of the low to medium gem quality ones in numerous other areas (Fig. 2.12).

Fig. 2.11 Gold ring with a larger triangular shaped table cut sapphire presumably from Sri Lanka (bigger dimension: 12 mm) and 12 smaller table cut rubies presumably from Burma (not tested by the authors); late sixteenth century. Ashmolean Museum, Oxford, England inv. WA1899.CDEF. F491. (Photo by Hugo Miguel Crespo)

Fig. 2.12 Rough corundum in host rocks from Longido (Kenya). Upper rock is 3 cm in length. (Photo: Stefanos Karampelas/LFG)

2.3 Turquoise and Lapis Lazuli

Turquoise and lapis lazuli are two of the oldest used gems. Turquoise ($Cu(Al,Fe^{+3})_6$ $(PO_4)_4(OH)_8 \cdot 4H_2O$), a blue to green opaque cryptocrystalline (massive) gem, almost never appears as single crystal (Fig. 2.13). It was believed that turquoise from Sinai

2.3 Turquoise and Lapis Lazuli

Fig. 2.13 A rough (right sample) and a polished turquoise (left sample). The rough sample is 2 cm in length. (Photo: Stefanos Karampelas/LFG)

Peninsula (Egypt) was used since 3200 BCE, and some researchers believe that it was used even before that time (Salanne 2009). Similarly, turquoise from Nishapur, Iran (formerly known as Persia), was used 2000 years ago, however some believe that it has been used even before that period. Currently, in the Hubei Province, China, turquoise has been exploited. During excavations, several turquoise pieces sculptured in the form of animals have been found which date back to 1300 BCE, in addition to a turquoise necklace which was believed to have been made from Chinese turquoise and dates back BCE (Fuquan 1986; Lijian et al. 1998). Turquoise from Sarazm (Tajikistan) is probably also of historical significance and nowadays is a source of good quality gem material. It was believed that civilizations were settled nearby this area since 1500 BCE but no indications of ancient mining have been found yet. Turquoise deposits of New Mexico (Los Cerillos), and probably California, were mined by pre-Columbian Native Americans (Salanne 2009). Jewellery items adorned with turquoise were made later on (Fig. 2.14) and today turquoise is highly appreciated by several cultures (e.g., Middle East).

Lapis lazuli is a rock (i.e. aggregate of several minerals) which contains principally lazurite as well as calcite and pyrite (Figs. 2.15 and 2.16). Sometimes it might also contain diopside, amphibole, feldspar and mica. The most famous mines of lapis lazuli are in Sar-e-Sang, Badakshan, today Afghanistan (Wyart et al. 1981). It was of great importance during the Bactrian times and was appreciated by Marco Polo (around the fourteenth century). These mines were worked for more than 6000 years and are believed to be the world's oldest known commercial gem source (Webster and Anderson 1983). Today important amounts are produced west of Lake Baikal in Russia, in the Andes (Chile) as well as in Mongolia, Burma and less from USA, Canada and Italy. Necklaces made of lapis-lazuli have been found in Neolithic tombs of Mauretania and Caucasus. Mesopotamia was the commercial centre for this gem material which was extensively used for making seals, necklaces and statues in the Ur's royal cemeteries (<3000 BCE). Objects made of lapis lazuli were also found in Ancient Egypt. It is believed that only the Sar-e-Sang mine was producing the high-end material and this material was transferred far from the source since antiquity (Wyart et al. 1981). In addition, it was also used as pigment and was widely used in Europe much later, *ca.* sixth century CE.

Fig. 2.14 Upper part of a gold ring with 16 turquoise beads presumably from Iran, 4 table cut rubies presumably from Burma and a table cut diamond (centre; about 2 mm long) presumably from India (not tested by the authors); seventeenth century. Ashmolean Museum, Oxford, England. inv. WA1899.CDEF.F489. (Photo by Hugo Miguel Crespo)

Fig. 2.15 Rough samples of lapis lazuli. Left sample is 7.8 cm in length. (Photo: Stefanos Karampelas/LFG)

Fig. 2.16 Polished gems of lapis lazuli. Left sample is 2.7 cm in diameter. (Photo: Stefanos Karampelas/LFG)

2.4 "Jade"

'Jade' is a trade name that for gemmologists refers to two virtually monomineralic rocks: 'nephrite jade' and 'jadeite jade'. 'Nephrite jade' is a microcrystalline to cryptocrystalline amphibolitic rock consisting of Ca-, Mg- and Fe-rich amphibole minerals from the solid solution series tremolite-ferroactinolite $(Ca_2(Mg,Fe)_5Si_8O_{22}(OH)_2)$, whereas 'jadeite jade' is a fine grained pyroxenitic rock called jadeitite. Jadeitite consists of at least 90% in volume of pyroxene, with the average pyroxene containing at least 90% in mole of sodic pyroxene jadeite $(NaAlSi_2O_6)$. For archaeologists, the term 'jade' has a broader meaning, including more rocks than in gemmology, e.g. fine-grained eclogites, serpentinites, nephrites, etc. which were also sometimes considered as 'jades' or mentioned as 'greenstones'. Additionally, "jade" has a different meaning in some languages; e.g., in Chinese the equivalent to "jade" which is the pictogram 玉 (*yü*) includes at least 20 rocks (Soubra 1999; Harlow et al. 2014).

Jade-like stones were used to make tools since the Paleolithic (prior 35,000 BCE) and only much later as gems (Fig. 2.17). "Jadeite-jade" has been used since the Neolithic in East Asia and Europe and later in the New World (D'Amico 2005; Harlow et al. 2014). In Japan "jadeite-jade" objects were known from the Jōmon period (*ca.* 3000 BCE), Neolithic along Europe and in Mesoamerica (southern Mexico through Guatemala to Honduras and Nicaragua) objects were found since the Early Formative period, *ca.* 1500–1000 BCE (Harlow et al. 2014). The important sources of "jadeite-jade" were at the Western Alps (Italy) for Europe in the middle of the Motagua Valley (Guatemala) for America. There is always the potential for small, as yet undiscovered, or exhausted, deposits to be an archaeological source. Potential sources of pre-Columbian "jadeite-jade" were recently found in

Fig. 2.17 A neolithic axe head made of jadeite (excavated at Lüscherz, Bern, Switzerland); 14 cm in the longest dimension. (Photo: Swiss National Museum)

Cuba and Dominican Republic (Harlow et al. 2014). "Jadeite jade" seems to have appeared more in China during the thirteenth century, and was believed that it started from the Qing Dynasty (the seventeenth century) in China, where the Chinese switched from nephrite jade to jadeite jade at a time when the Chinese empire extended into the Yunnan Province and the Kachin State of present-day Myanmar (Hughes et al. 2000). Nowadays, several mines are active with those in Burma being the most important. Other pyroxenitic rocks (omphacitite) were also considered jade ("omphacite-jade") as well as omphacite- and jadeite- rich pyroxenitic rocks ("omphacite-jadeite jade"). Some items of archaeological importance were identified to be "omphacite-jade" (Coccato et al. 2014).

"Nephrite-jade" is more common than "jadeite-jade". It has been used as far back as "jadeite-jade", probably longer than 7000 years in China and other places. "Nephrite-jade" objects from Bulgaria have been recently re-determined and dated back to 7000–6000 BCE (Kostov 2010). Later nephrite jade was used by Maoris and also the North American natives. "Nephrite-jade" was found in innumerable small deposits. Presently, some of the most important nephrite deposits are located in China and Canada.

2.5 Diamonds

Diamonds started to be used later than various other gems. Some people believed that they have been used since the fourth century BCE and a reference was found in Pliny *ca.* first century CE. This relatively late use was due to its rare occurrence and the fact that it was considered to be the hardest of gem materials found. Until the sixth century, India was the world's only source of gem diamonds (Levinson et al. 1992; Harlow 1998). Some researchers believed that diamonds from Kalimantan, Borneo, were mined since the sixth century CE while other researchers state that this was since the tenth century CE, some fourteenth century CE and some other not before the sixteenth century CE (Spencer et al. 1988). It was believed that these diamonds were traded locally or were included in the Indian production before

Fig. 2.18 Two rough diamond crystals on the right and two round brilliant cut diamonds on the left. The rough diamond crystals are about 5 mm in length. (Photo: Bérengère Meslin Sainte Beuve/LFG)

being exported to Europe (Stachel 2014). Diamonds were mostly appreciated by Indians and in around fourteenth century found their way outside. In parallel with the development of the gem cutting industry, diamonds were becoming fashionable in Europe (Klein 2005). An important source of diamonds in the first quarter of the eighteenth century was Brazil and at the end of the nineteenth century diamonds were found in South Africa. In Russia, Guyana and Australia diamonds were also found during the nineteenth century but larger scale mining only took place in the twentieth century along with several areas in Africa, Canada and in Venezuela (Levinson et al. 1992). Today diamond, principally round brilliant cut, is the most popular gem world-wide (Fig. 2.18).

2.6 Emerald and Other Beryls

Emerald is the green coloured variety of beryl ($Be_3Al_2Si_6O_{18}$). The earliest known emerald mine is located at Wadi Sikait in Egypt's Eastern Desert (Harrell 2004). It was believed that these mines were exploited before 1500 BCE (Gonthier 1998), and that this stone could have existed even before, nearly 4000 BCE. The mines in Egypt, a.k.a. Cleopatra's emerald mines, were worked from at least 2000 BCE (Webster and Anderson 1983). However, with the exception of one emerald bead which was dated to the predynastic times, emeralds appeared from the fourth century BCE and onwards (Jennings et al. 1993). The most active period probably occurred in the last first to the sixth century CE (Groat et al. 2014). Mining also took place at several other sites within a 15 km radius of Wadi Sikait from mid-sixth century CE onwards (Harrell 2004). Emeralds from Austria (Habachtal) may have been mined in antiquity by the Romans or even the Celts (Gubelin 1956; Gonthier 1998). Studies on ancient jewellery pieces have confirmed that emeralds from Egypt and Austria were used since the third century CE (Calligaro et al. 2000; Giuliani et al. 2001). The emeralds found in Austria and Egypt were relatively large with a pronounced green colour but with low transparency; they were mostly translucent to opaque (Rondeau 2003). Studies on an emerald mounted in a Gallic-Roman earring suggested that it was from an occurrence in Swat Valley, Pakistan (Giuliani et al. 2000). Emeralds from Ural Mountains (Russia) and Panjshir valley (Afghanistan) were presumably used since the antiquity; stones from these mines have not been identified until now in ancient jewellery (Giuliani et al. 2001). On the other hand, emeralds from Afghanistan were found in treasures dated 1000 CE (Schwarz and Pardieu 2009). Additionally, small emerald occurrences could have been the source of emeralds in antiquity.

The emeralds in antiquity were used either engraved, simply polishing the rough crystals or as cabochons. Emeralds were described in both Theophrastus and Pliny the Elder; where they mentioned several origins. For instance, Theophrastus wrote about Cyprus, Bactrian or Laconian origins and Pliny mentioned about 12 different sources, including Scythia, India. It is suggested that they meant different green coloured gems (such as malachite from Cyprus, green turquoise or variscite from Laconia) and not emeralds (Strack and Kostov 2010).

The main change in the use of emeralds occurred in the middle of the sixteenth century when the Spaniards exploited the emerald mines in Chivor, Colombia (Giuliani et al. 2000). After then, emeralds of high transparency were present in jewellery (Fig. 2.19). However, it was believed that emeralds from Colombia were used from well before (500 BCE–200 CE; emerald figurine from an "unknown" excavation or earlier than 1000 CE; Whittington et al. 1998). Emeralds from Russia (Ural Mountains) and Brazil (Brumado in Goias) were found at around 1830 and 1910 but were exploited at a later stage (Groat et al. 2014). As mentioned above, emeralds from Ural Mountains presumably used since the antiquity, however neither stones from these mines have been identified until now in ancient jewellery nor traces of ancient mining activity was found. Relatively significant quantities of emeralds from an occurrence in Norway (Byrud) were exploited at the end of the nineteenth century (Rondeau et al. 2008). In the course of the twentieth century

Fig. 2.19 A gold brooch (6.3 cm at its longest dimension) adorned with 25 emeralds presenting inclusions similar to those observed in emeralds from Colombia (see Fig. 3.3), second half of seventeenth century (not tested by the authors). (Photo by Carlos Pombo Monteiro; © Fundação Eugénio de Almeida/Archidioceses of Évora, Portugal)

2.6 Emerald and Other Beryls

Fig. 2.20 Emeralds hosted in different host rocks from various mines. Left rock is from Colombia and right rock from Brazil. The right rock is 4.5 cm in the longest dimension. (Photo: Stefanos Karampelas/LFG)

Fig. 2.21 Gemstones of the beryl family (clockwise from left): Red beryl (faceted and rough), emerald (green), heliodor (yellow), aquamarine (light blue), morganite (light yellow) and pezzottaite (pink). The rough crystal on the left is 11 mm in length. (Photo: Lore Kiefert)

several important emerald mines were discovered in Africa as well as in South America (Fig. 2.20).

Aquamarine is a blue to greenish-blue, sometimes very light, coloured variety of beryl; one of which was engraved from Hellenistic period onwards (Richter 1956). Some gems in the treasure of Guarrazar (Spain, mid-sixth century CE) were thought to be aquamarine (Rapp 2002a). The exact origin of the ancient aquamarines is not known, but it was believed that they are derived from Sri Lanka, Burma, India and Russia (Ural Mountains). Afghanistan and Pakistan along with these small occurrences were also probable sources. In Russia, aquamarine was systematically exploited from the eighteenth century and at the end of the nineteenth century in Brazil. Nowadays, several occurrences were found in Africa. Heliodor is the brownish- to orangey-yellow to yellow and morganite is the pink coloured beryl variety; other coloured beryl varieties such as red beryl and pezzottaite (caesium and lithium rich mineral closely related to beryl) are also researched today (Fig. 2.21) and are probably used since a long time, but were usually confused with other gems, however references on this matter are scarce.

2.7 Garnets

Garnet is a group of minerals with a general chemical formula $X_3Y_2(SiO_4)_3$ where X: Ca, Mg, Fe or Mn (divalent cations) and Y: Al, Fe, Cr (trivalent cations). Pyralspite garnets are those with Al in Y site and includes pyrope ($Mg_3Al_2(SiO_4)_3$), almandine ($Fe_3Al_2(SiO_4)_3$) and spessartite ($Mn_3Al_2(SiO_4)_3$). Ugrandite garnets are those with Ca in X site and include uvarovite ($Ca_3Cr_2(SiO_4)_3$), grossular ($Ca_3Al_2(SiO_4)_3$) and andradite ($Ca_3Fe_2(SiO_4)_3$). Natural garnets usually are mixtures (solid solution) of the end members in various proportions.

Red, purplish and pinkish coloured garnets, as well as pink to purple, were used sporadically in Egypt *ca.* 4000–2000 BCE and in the Middle East *ca.* 2000–1000 BCE (Thoresen and Schmetzer 2013). They started to be more popular from about 300 BCE, and the period between 300 BCE and 700 CE was referred to by some as the "garnet millennium" (Adams 2011). The same terms were used for almost all red stones; with anthrax and carbunculus the terms used by Theophrastus and Pliny respectively. However, the majority of red stones that dated before the ninth century AD were garnets, not spinel or ruby. Garnet was shaped in cabochons and cameos for personal ornaments, engraved as ring stones and, at the end of this time frame, polished in flat plates that were assembled in cloisonné cellwork (Fig. 2.22). After this millennium of intensive exploitation, garnets continued to be used and re-used,

Fig. 2.22 Longobard gold pendant with a cloisonné including slices of almandine garnets coming from Lodi Vecchio (Italy). Length of the pendant approximately 3 cm. (Photo: Danilo Bersani. Reproduced with the permission of "Ministero per i Beni e le Attività Culturali e per il Turismo – Soprintendenza Archeologia, Belle Arti e Paesaggio per le Province di Cremona, Lodi e Mantova, Italy")

Fig. 2.23 A silver and gold brooch (sized 2.6 cm in its longest dimension) mounted "en tremblant" with two hessonite garnets, natural saltwater pearls and 11 rose cut diamonds (not tested by the authors). Portugal, late eighteenth century. (Photo by Carlos Pombo Monteiro; © Fundação Eugénio de Almeida/Archidioceses of Évora, Portugal)

notably on Early Medieval liturgical objects, but were not a dominant feature of gemstone jewellery until the upsurge of production of pyrope garnet from Bohemian mines, when they started to become popular once again. Orange coloured grossular garnets (Fig. 2.23), principally from India and Sri Lanka, were also used in antique jewellery, sometimes under the name "hyacinth" or "hessonite", used also for similarly coloured zircons.

Different authors have suggested various categories of garnets found in antique jewellery; all being of the pyralspite garnet group (some examples can be found at Rösch et al. 1997; Farges 1998; Calligaro et al. 2006–2007; Horváth and Bendö 2011; Thoresen and Schmetzer 2013). The most recent suggestion is Mn-poor/Cr-free almandine, Mn-rich/Cr-bearing almandine, Ca-rich almandine, intermediate pyrope-almandine, Cr-poor pyrope and Cr-rich pyrope. Ancient Cr-rich pyrope could have been from Bohemia, which is still mined today in the Czech Republic. These stones are relatively small (Schlüter and Weitschat 1991; Hyrsl 2001a). Various localities in India (especially in Rajasthan and south India) have been claimed to be possible origins for two types of almandine garnet and Sri Lanka has been postulated as a source of intermediate pyrope-almandine garnet. However, some researchers believe some of the antique garnets could have been sourced from small occurrences in Austria for almandine garnets, the Czech Republic for almandine and pyrope-almandine garnets, without excluding some sources in Africa, Scandinavia etc. (Hyrsl 2001a; Adams et al. 2011; Fritsch et al. 2010; Thoresen and Schmetzer 2013).

2.8 Coral, Ivory and Amber

Coral, amber and ivory, along with pearls, are also considered organogenic (biogenic) gems. The term 'coral' refers to many Anthozoa and to some Hydrozoa (marine polyps-cnidarian invertebrates) that develop colonies forming a common calcareous skeleton, sometimes building huge reefs in the shallow seawater. Only the skeleton, made principally of calcite (and/or rarely aragonite), organic matter and water, is used as gem material. Some corals are made almost only of organic matter. However, the vast majority of corals (around 7300 species), are not used in jewellery. The most corals used as gems are produced in eight species of marine polyps belonging to the family Corallidae of the Anthozoa class, that live in relatively deep water. The most valuable variety of coral are pink to red in colour. Coral was used since the early Neolithic times in Europe, some coral artefacts were found in Sumerian and Egypt from around 3000 BCE, and from far Eastern cultures since around 1000 BCE (Smith et al. 2007). Corals from *Corallium* sp. are the most used ones as gem, with the Eastern Mediterranean Sea as the most probable source (Fig. 2.24). The waters off today's Japan, China and Taiwan are also a possible sources of such corals. They were used through all periods, also from Celts, during Middle Age and in ecclesiastic items. Coral was mostly fashioned as beads but some cameos were also found.

Ivory is a substance that comprises of teeth or teeth modifications such as the tusks of mammals. The most valuable ivory which is used as gem material comes from the mammoth (*Mammuthus* sp.) and elephants (*Elephas* sp. -a.k.a. Asian elephant- and *Loxodonta* sp. -a.k.a. African elephant-; Fig. 2.25). Evidence of 30,000-year-old engraved mammoth and elephant ivory items were found in several caves (e.g., Vogelherd Cave -Germany-, Madeleine cave -France-) in central Europe (Kunz 1916; Webster and Anderson 1983). All different civilizations used ivory with important pieces produced during Minoan times (fifteenth century BCE), and ivory treasures from Tutankhamun tomb (fourteenth century BCE) as well as others were found (Caskey 1915; Kunz 1916; Webster and Anderson 1983). Most of the ancient ivory is believed to come from North Africa and from today's Syria where the elephant population was reduced to extinction 2000 years ago. Throughout the ages ivory was used in various forms, with great importance in Japan and China as well as Sri Lanka and India due to its symbolic value.

Amber is a time hardened, fossilized, tree resin. Not all tree resin can become amber. In general, resins that fossilize to amber are secreted by trees in the families of *Araucariaceae* and *Fabaceae* (Abduriyim et al. 2009). The exact tree responsible for the origin of Baltic amber for example is still under discussion (Wolfe et al. 2009). The use of amber from the Baltic Sea dates as far as the early Neolithic civilizations (de Navarro 1925; Grimaldi 2009). The Amber Road was one of the most important trade roads (Fig. 2.26). It was believed that as early as the sixteenth century BCE amber was moved from Northern Europe to the Mediterranean Sea (Harding et al. 1974). Amber was also found in the Tutankhamun tomb and was used during the classical times. During the Middle Ages it was also widely used with the Baltic coast (e.g. Gdansk and Kaliningrad) as main source (Fig. 2.27).

2.8 Coral, Ivory and Amber

Fig. 2.24 Exceptional set consisting of a necklace, a pair of earrings, a brooch, a bracelet and a tiara, hand-carved in the early nineteenth century in Torre del Greco from corals belonging to *Corallium rubrum*, reportedly off Sciacca, south of Sicily, Italy. This set was commissioned by Joaquin-Napoleón Murat, king of Naples (from 1808 to 1815) and his wife Caroline Bonaparte for a gift to Empress Josephine (Josephine of Beauharnais), Napoleon I's wife until January 1810 when Napoleon Bonaparte arranged for the nullification of the marriage. Because of this divorce, the necklace remained undelivered with the manufacturer and resides now at the Museo del Corallo – Colezzione Liverino, in Torre del Greco, near Naples, Italy (not tested by the authors). Brooch is 8 cm long. (Photo © Museo del Corallo – Colleziones Liverino)

Fig. 2.25 Brooch from elephant ivory. The biggest dimension is 4.5 cm. (Photo: Stefanos Karampelas/LFG)

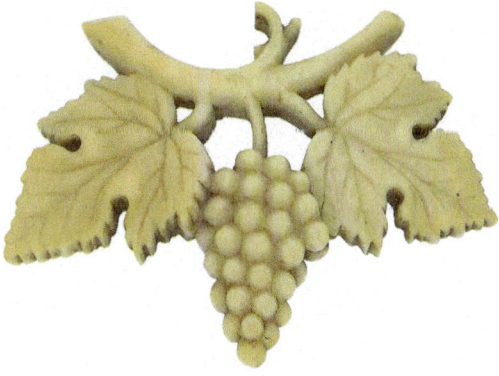

Fig. 2.26 A rosary (120 cm long) made of large amber beads. Portugal, seventeenth/eighteenth century (not tested by the authors). (Photo by Carlos Pombo Monteiro; © Fundação Eugénio de Almeida/Archidioceses of Évora, Portugal)

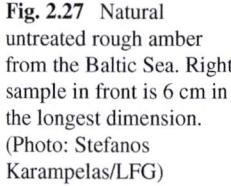

Fig. 2.27 Natural untreated rough amber from the Baltic Sea. Right sample in front is 6 cm in the longest dimension. (Photo: Stefanos Karampelas/LFG)

Fig. 2.28 Photomicrographs (photographs through a microscope) of a natural amber with an insect (left) and part of a plant (right). Field of view: 5.5 mm (left) and 1.4 mm (right). (Photomicrographs: Ugo Hennebois/LFG)

Some collectors look for amber as inclusions of insects and plants from the past can be stored and they are of high geological and paleontological significance (Fig. 2.28). Another important source of amber is Burma. Copal, a younger tree resin, was also sometimes used in the past by some civilizations (e.g., in America) and today is widespread in the market.

2.9 Tourmaline, Spinel, Peridot, Topaz, Chrysoberyl and Zircon

Tourmaline was firstly used in antiquity after the conquest of Alexander the Great in the East; with Afghanistan, Pakistan, India and Sri Lanka been possible sources (Rapp 2002a). It is the gem with the biggest variety of colours, even within individual crystals (Fig. 2.29; Pezzotta and Laurs 2011). However, the most used tourmaline in the old times was the purple to pink- to red tourmaline (some of them a.k.a. rubellite), green tourmaline and black tourmaline (schorl). Tourmaline belongs to the tourmaline group of minerals with variable chemistry (Henry et al. 2011). Most gem quality tourmaline belongs to the solid solution series with (fluor)-elbaite, (fluor)-liddicoatite, rossmanite, dravite and schorl as end members. Tourmaline was recognized as a new type of gem, having been unrecognized and mixed with other gems or misidentified till then, at the beginning of the eighteenth century (Pezzotta and Laurs 2011). Studies of the St. Wenceslas crown in Prague (made mid-fourteenth century CE) show that one of the large sized red stones is a tourmaline (Hyrsl and Neumanova 1999). Green tourmaline was also confused with other green stones.

Fig. 2.29 Tourmaline occurs in a variety of colours. The black rough crystal in the back is 35 mm in length. (Photo: Lore Kiefert)

Fig. 2.30 Gems from the spinel group (length of the rough pink crystal in the back: 11.4 mm). (Photo: Tanja Jordi, GGL)

Fig. 2.31 Peridot from Myanmar. Length of the longer gemstone on the top left: 10.2 mm. (Photo: Lore Kiefert)

Brazil became an important source of tourmaline after the seventeenth century; important deposits were also found in Italy (e.g. Elba Island) and California (USA). "Paraiba-type" tourmaline ("neon" blue to green Cu-and Mn- bearing Na- or Ca-rich tourmaline) was discovered in the late 1980s in Brazil and later on in Mozambique and Nigeria and is distinctly different to blue tourmaline coloured by iron.

Spinel ($MgAl_2O_4$) is another gem which was probably included under the terms anthrax and carbunculus, the terms used by Theophrastus and Pliny respectively for the red stones. It was accepted as a different mineral in the mid-eighteenth century. Pink-to-red coloured spinel was used in royal jewellery; with the famous spinel known as "Black Prince's ruby" of the Imperial State Crown (mid-fourteenth century AD) and "Timur ruby" (Webster and Anderson 1983). Other coloured spinel varieties were also found with the blue and green as the most valuable (Fig. 2.30). The most important pink-to-red historic large spinels were found in the Kuh-i-Lal (Badakhshan) mines situated in nowadays Tajikistan, with Sri Lanka also being a possible, but less common, old spinel mining area.

Peridot is a term used by gemmologists for the gem quality yellowish-green to green olivine solid solution series (($Mg,Fe)_2SiO_4$); closer to Mg-rich (forsterite) member (Fig. 2.31). In old texts this gem was called topazios as well as chrysolite (Gubelin 1981), but the latter is no longer used. Topaz is another gem producing mineral with the chemical formula $Al_2SiO_4(F,OH)_2$. Topazios was also used to refer

2.9 Tourmaline, Spinel, Peridot, Topaz, Chrysoberyl and Zircon

Fig. 2.32 Various coloured topaz. Length of the light blue topaz crystal on the top left: 25 mm. (Photo: Lore Kiefert)

to the source of peridot, which is the island of Zabargad, today known as St John Island, in the Red Sea. It is the oldest known source of gem peridot, probably even from before 1000 BCE by Egyptians, but used only rarely. It has been used in prestigious items, probably as emeralds, as well as in several ecclesiastic items in Europe after the twelfth century (Hyrsl 2012).

True topaz was used from the eighteenth century onwards (Webster and Anderson 1983). Gem quality topaz comes in a variety of colours, the most popular of which are pink to reddish to purple, yellow to orange as well as blue and green (Fig. 2.32). The most demanded topaz is that of "imperial" colour, which ranges from pinkish- to reddish-orange. The main mines are situated in Brazil and Russia, with an occurrence found in Germany, but also from areas such as Afghanistan, Pakistan, Sri Lanka etc. (Rapp 2002a). Some topaz from Brazil were used for the crown jewels at the end of the eighteenth century in France.

Chrysoberyl ($BeAl_2O_4$) can be yellow to brown and was used since antiquity, probably before the Roman Empire, most likely also under the name of chrysolite (Fig. 2.33). It can also present chatoyancy (or cat's eye effect), which is an optical effect due to the orientation of inclusions within a stone. Other gems can also present this effect. Most of chrysoberyl from the antiquity were most likely derived from India and Sri Lanka. Some chrysoberyl display a colour change, i.e. different colours under daylight and incandescent light. This kind of chrysoberyl is called alexandrite, green to red colour change is the most demanded, with the most important occurrences found in Russia and Brazil at the beginning of the nineteenth century as well as some in Sri Lanka.

Zircon ($ZrSiO_4$) of orangey-brown to brown colour was used since about 2000 years and it was believed that this was sourced from India (Rapp 2002a). A long list of names was used for this gem, with hyacinth as the most common (Rouille 2010). The other colour varieties (e.g. colourless and blue) started to be used later (Fig. 2.34).

Fig. 2.33 Three faceted and one rough natural untreated chrysoberyl from Sri Lanka. The rough sample in front is 1 cm in length. (Photo: Stefanos Karampelas/LFG)

Fig. 2.34 Zircon comes in a range of colours, here a small selection. Length of the octagonal blue zircon: 8.8 mm. (Photo: Lore Kiefert)

2.10 Silica Gems

Silica (SiO$_2$) gems were widely used during antiquity. Different varieties belonged to these gems which usually created nomenclature issues; the appearance under the microscope and colour are the most important characteristics. Crystalline varieties (Fig. 2.35), single mainly transparent crystals, include different coloured quartz such as rock crystal (or simply quartz), amethyst (purple), citrine (yellow), ametrine (amethyst and citrine in the same crystal), rose quartz (pink), smoky quartz (grey to brown) and milky quartz (white) the most common. Some of them can present chatoyancy or asterism. Some quartz crystals coloured by their inclusions were also used in antiquity (e.g., aventurine, prasiolite).

In the ancient Egypt (before 2000 BCE) quartz was used to make scarab seals as well as in jewellery during the Minoan and Mycenaean period (Gray 1983). In the Far East, as well as in the Middle East and South America, these stones were also used during the same time period (Rapp 2002a, b). References on these gems can also be found in Theophrastus' and Pliny's works. Ancient sources were reported from various localities, with Sri Lanka, India, Burma and areas in the Middle East the most commonly cited. Small, local occurrences cannot be excluded though; the Alpine region is believed to be one. High quality amethyst was found in the Ural Mountains (Russia), and was used for the crown jewels at the end of the eighteenth century. Important quantities of quartz were found in Brazil after the New World discovery and also in Africa in the course of the twentieth century.

Cryptocrystalline-massive silica (SiO$_2$) gems were also widely used during antiquity as seal stones, cameo and intaglio. These gems are mainly translucent to opaque and the crystals can only be resolved under the microscope. The term chalcedony covers all cryptocrystalline varieties of silica (Fig. 2.36). Varieties of chalcedony were favoured at least from 2000 BCE and used to make cylinder seals and scarabs (Gray 1983). Different colours and motifs had different names. The green coloured variety was one of the most valuable in antiquity until today (Hyrsl 1999; Shen et al. 2006; Lüle 2011). Chrysoprase is a greenish coloured chalcedony coloured by "nickel-rich" green inclusions, chrysocolla chalcedony (or "gem sil-

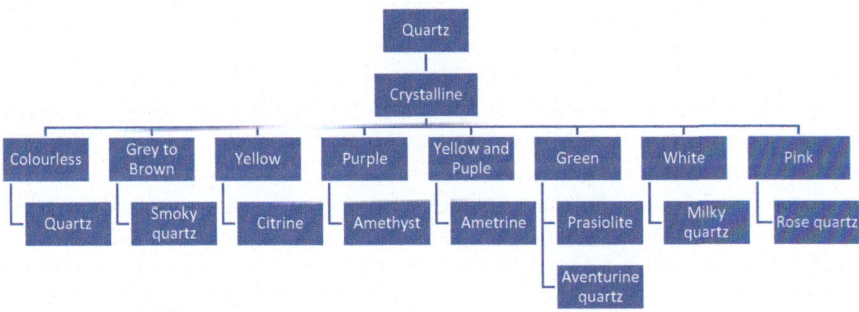

Fig. 2.35 Varieties for crystalline quartz following their colouration

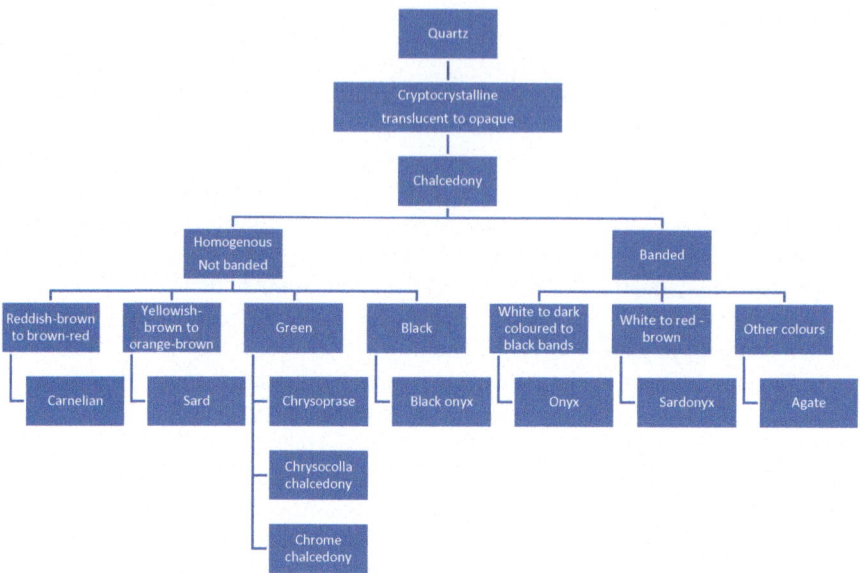

Fig. 2.36 Varieties for cryptocrystalline silica following their appearance

ica") is greenish coloured chalcedony coloured by chrysocolla inclusions and chrome chalcedony coloured by chromium. Egyptian artefacts with chrysoprase from predynastic period were reported. It is not known where the chrysoprase comes from; the oldest known significant deposit was exploited in the early fifteenth century near Szklary (nowadays Poland; Hyrsl 1999). It has been found that chrome chalcedony is present in some Roman intaglios mainly from first to third century CE; however, the exact source is not yet known (Hyrsl 1999; Lüle 2011). Several assumptions have been made which emphasized that this material could have come from Africa (Zimbabwe) or nowadays Turkey. Carnelian is commonly translucent reddish-brown to brown-red coloured chalcedony and sard is the yellowish-brown to orange-brown counterpart.

Jasper is opaque brown to red coloured and it is not made only of chalcedony but also of other minerals and it was firstly used by the Assyrians. The earlier scarabs were frequently made of jasper due to its availability. The Eastern desert of Egypt was mentioned as one of the antique origins of carnelian and sard; artefacts dated from sixth to seventh millennium BCE and were found in present India and Pakistan (Rapp 2002a). Artefacts in Mesopotamia and India from 3rd millennium BCE was believed to be made of material found in India.

Banded translucent chalcedony with various colours, usually curved, is called agate (there are also other types of agate such as iris and moss). If the bands are straight white and dark coloured to black they are referred to as onyx (black onyx is without banding and nicolo is called the onyx when its white-gray part is mounted on the top -commonly engraved- and the dark blue-black part facing down) and if the bands are straight white and brown to red it is called sardonyx. Magnificent cameos and intaglios were made of these gem materials from the third century AD.

Opal is a hydrated nanocrystallized to amorphous silica (SiO_2 nH_2O). Common opal is the gem which does not show play of colour while precious opal does. Opal can have different colours such as white, pink, orange to brown ("fire" opal), black, green etc. Opal was used as gem since the Roman times (Webster and Anderson 1983). Opal deposits at Dubnil, nowadays in Slovakia (a.k.a. "Hungarian" opal) were the unique source of gem opal in Europe till the nineteenth century (Rondeau et al. 2004). Mexican "fire" opal deposits were probably known to the Aztecs since the thirteenth century AD (Webster and Anderson 1983). Opal deposits in Australia were found in the mid-nineteenth century at about the time other deposits were found in South America.

References

Abduriyim A, Kimura H, Yokoyama Y, Nakazono H, Wakatsuki M, Shimizu T, Tansho M, Ohki S (2009) Characterization of 'green amber' with infrared and nuclear magnetic resonance spectroscopy. Gems Gemol 45:158–177

Adams N (2011) The garnet millennium: the role of seal stones in garnet studies. In: 'Gems of heaven': recent research on engraved gemstones in late antiquity c. ad 200–600, vol 177. British Museum Research Publication, London, pp 10–24

Adams N, Lüle C, Passmore E (2011) Lithóis Indikois: preliminary characterization of garnet seal stones from Central and South Asia. In: 'Gems of heaven': recent research on engraved gemstones in late antiquity c. ad 200–600, vol 177. British Museum Research Publication, London, pp 25–38

Ainis AF, Fujita H, Vellanoweth RL (2019) The antiquity of pearling in the Americas: pearl modification beginning at least 8500 years ago in Baja California Sur, Mexico. Latin American Antiquity (in press)

Atkinson D, Kothavala RZ (1981) Kashmir sapphires. Gems Gemol 19:64–76

Bari H (2007) La perle rose: Trésor des Caraïbes. Skira, Milano, 176 pp

Bari H, Lam D (2009) Pearls. Skira, Milano, 336 pp

Bowersox GW, Foord EE, Laurs BM, Shigley JE, Smith CP (2000) Ruby and sapphire from Jegdalek, Afghanistan. Gems Gemol 36:110–126

Calligaro T, Dran JC, Poirot JP, Querre G, Salomon J, Zwaan JC (2000) PIXE/PIGE characterisation of emeralds using an external micro-beam. Nucl Inst Methods Phys Res B 161–163:769–774

Calligaro T, Perin P, Vallet F, Poirot JP (2006–2007) Contribution á l'étude des grenats mérovingiens (Basilique de saint-Denis et autres collections du musée d'Archéologie nationale, diverses collections publiques et objects de fouilles récentes). Antiquités Nationales 38:111–144

Carino M, Monteforte M (1995) History of pearling in La Paz Bay, South Baja California. Gems Gemol 31:88–105

Caskey LD (1915) A chryselephantine statuette of the Chetan snake goddess. Am J Archaeol 19:237–249

Charpentier V, Phillips CS, Méry S (2012) Pearl fishing in the ancient world: 7500 BP. Arab Archaeol Epigr 23:1–6

Coccato A, Karampelas S, Wörle M, van Willingen S, Pétrequin P (2014) Gem quality and archeological green "jadeite jade" vs "omphacite jade". J Raman Spectrosc 45:1260–1265

D'Amico C (2005) Neolithic 'greenstone' axe blades from NorthWestern Italy across Europe: a first petrographic comparison. Archaeometry 47:235–252

de Navarro JM (1925) Prehistoric routes between Northern Europe and Italy defined by the amber trade. Geogr J 66:481–503

Dirlam DM, Misiorowski EB, Thomas SA (1985) Pearl fashion through the ages. Gems Gemol 21:63–78

Donkin RA (1998) Beyond price: pearls and pearl-fishing, origins to the age of discoveries. American Philosophical Society, Philadelphia, 448 pp

Farges F (1998) Mineralogy of the Louvres Merovingian garnet cloisonné jewelry: origins of the gems of the first kings of France. Am Mineral 83:323–330

Fritsch E, Ionescu C, Simon V, Nagy S, Nagy-Pora K, Rotea M (2010) 5th century garnet jewelry from Romania. Gems Gemol 46:316–318

Fujita H, Caceres-Martinez C, Ainis AF (2017) Pearl ornaments from the Covacha Babisuri site; Espiritu Santo Island, Baja California Sur, Mexico. Pac Coast Archaeol Soc Q 53:63–86

Fuquan W (1986) A gemological study of turquoise in China. Gems Gemol 22:35–37

Giuliani G, Chaussidon M, Schubnel HJ, Piat D, Rollion-Bard C, France-Lanord C, Giard D, de Narvaez D, Rondeau B (2000) Oxygen isotopes and emerald trade routes since antiquity. Science 287:631–633

Giuliani G, Chaussidon M, France-Lanord C, Savay Guerraz H, Chiappero PJ, Schubnel HJ, Gavrilenko E, Schwarz D (2001) L'exploitation des mines d'émeraude d'Autriche et de la Haute Egypte à l'époque Gallo-Romaine: mythe ou réalité? Revue de Gemmologie AFG 143:20–24

Giuliani G, Ohnenstetter D, Fallick AE, Groat L, Fagan AJ (2014) The geology and genesis of gem corundum deposits. In: Geology of Gem deposits, vol 44. Mineralogical Association of Canada, Québec, pp 29–112

Goebel M, Dirlam DM (1989) Polynesian black pearls. Gems Gemol 25:130–148

Gonthier E (1998) The symbolic representation of famous emeralds in history. In: The emerald. Association Francaise de Gemmologie, Special Issue, pp 27–32

Gray FL (1983) Engraved gems: a historical perspective. Gems Gemol 19:191–201

Grimaldi D (2009) Pushing back amber production. Science 326:31–32

Groat L, Giuliani G, Marshall D, Turner D (2014) Emeralds. In: Geology of Gem deposits, vol 44. Mineralogical Association of Canada, Québec, pp 135–174

Gubelin E (1956) The emerald from Habachtal. Gems Gemol 8:295–309

Gubelin E (1981) Zabargad: the ancient peridot island in the red sea. Gems Gemol 17:2–8

Harding A, Hughes-Brock H, Beck CW (1974) Amber in the Mycenaean world. Annu Br Sch Athens 69:145–172

Harlow GE (1998) Following the history of diamonds. In: American Museum of Natural History (AMNH) (ed) The nature of diamonds. Cambridge University Press, Cambridge, pp 116–135

Harlow GE, Sorensen SS, Sisson VB, Shi G (2014) The geology of jade deposits. In: Geology of Gem deposits, vol 44. Mineralogical Association of Canada, Québec, pp 305–374

Harrell JA (2004) Archaeological geology of the world's first emerald mine. Geosci Can 31:69–76

Henry DJ, Novak M, Hawthorne FC, Ertl A, Dutrow BI, Uher P, Pezzotta F (2011) Nomenclature of the tourmaline supergroup minerals. Am Mineral 9:895–913

Horváth E, Bendö Z (2011) Provenance study on a collection of loose garnets from a Gepidic period grave in Northeast Hungary. Archeometriai Mühely 8:17–32

Hughes R, Thoresen L (2017) Chapter 1: history. In: Hughes R (ed) Ruby & Sapphire: a gemmologist's guide. RWH Publishing, Bangkok, pp 51–75

Hughes RW, Galibert O, Bosshart G, Ward F, Oo T, Smith M, Sun TT, Harlow GE (2000) Burmese jade: the inscrutable gem. Gems Gemol 36:2–25

Hyrsl J (1999) Chrome chalcedony – a review. J Gemmol 26:364–370

Hyrsl J (2001a) New gemmological study of large garnets of supposedly Czech origin. Gemmologie (Zeitschrift der Deutschen Gemmologischen Gesellschaft) 50:37–42

Hyrsl J (2001b) Sapphires and their imitations on medieval art objects. Gemmologie (Zeitschrift der Deutschen Gemmologischen Gesellschaft) 50:153–162

Hyrsl J (2012) Historical use of olivine – the origin of peridots in baroque period jewellery. Gemmologie (Zeitschrift der Deutschen Gemmologischen Gesellschaft) 61:35–42

Hyrsl J, Neumanova P (1999) A new gemmological study of the St. Wenceslas crown in Prague. Gemmologie (Zeitschrift der Deutschen Gemmologischen Gesellschaft) 48:29–36

Jennings RH, Kammerling RC, Kovaltchouk A, Calederon GP, El Baz MK, Koivula JI (1993) Emeralds and green beryls of upper Egypt. Gems Gemol 29:100–115

Klein G (2005) Faceting history: cutting diamonds and colored stones. Xlibris, Bloomington, 242 pp

Kostov RI (2010) Gem minerals and materials from the Neolithic and Chalcolithic periods in Bulgaria and their impact on the history of gemmology. In: Proceedings of the 19th CBGA congress, Bulgarian Academy of Sciences, pp 391–397

Kunz GF (1916) Ivory and the elephant in art, in archaeology and in science. Doubleday, Page and Co., New York, 527 pp

Kunz GF, Stevenson CH (1908) The book of the pearl. The Century Co., New York, 548 pp

Levinson AA, Gurney JJ, Kirkley MB (1992) Diamond sources and production: past, present and future. Gems Gemol 28:234–254

Lijian Q, Weixuan Y, Mingxin Y (1998) Turquoise from Hubei Province. J Gemmol 26:17–23

Lüle C (2011) Non-destructive gemmological test of the identification of ancient gems. In: Gems of heaven: recent research on engraved gemstones in late antiquity c. ad 200–600, vol 177. British Museum Research Publication, London, pp 1–3

Middleton JH (1891) The engraved gems of classical times. Cambridge University Press, Cambridge, 157 pp

Mocquet B (2003) Détermination de la nature pétrographique d'objets archéologiques du Musée Dobrée de Nantes. DUG Diploma thesis, University of Nantes, Nantes, France

Pezzotta F, Laurs BM (2011) Tourmaline: the kaleidoscopic gemstone. Elements 7:333–339

Rapp GR (2002a) Gemstones, sealstones, and ceremonial stones. In: Archaeomineralogy. Springer, Berlin/Heidelberg/New York, pp 91–120

Rapp GR (2002b) Abrasive, salt, shells and miscellaneous geological raw materials. In: Archaeomineralogy. Springer, Berlin/Heidelberg/New York, pp 219–241

Richter GMA (1956) Catalogue of engraved gems: Greek, Etruscan and Roman. L'Erma di Bretschneider, Rome, Italy, 143 pp

Rondeau B (2003) Matériaux gemmes de référence du Museum National D'Histoire Naturelle: exemples de valorisation scientifique d'une collection de minéralogie et gemmologie. PhD thesis, University of Nantes, Nantes, France

Rondeau B, Fritsch E, Guiraud M, Renac C (2004) Opals from Slovakia ("Hungarian" opals): a re-assessment of the conditions of formation. Eur J Mineral 16:789–799

Rondeau B, Fritsch E, Peucat JJ, Nordum FS, Groat L (2008) Characterization of Emeralds from a historical deposit: Byrud (Eidsvoll), Norway. Gems Gemol 44:108–122

Rösch RC, Hock R, Schüssler U, Yule P, Hannibal A (1997) Electron microprobe analysis and X-ray diffraction methods in archaeometry: investigations on ancient beads from Sultanate of Oman and from Sri Lanka. Eur J Mineral 9:763–783

Rouille M (2010) The zircon. DUG Diploma thesis, University of Nantes, Nantes, France

Salanne C (2009) Etude de la turquoise, de ses traitements et imitations. DUG Diploma thesis, University of Nantes, Nantes, France

Scarratt K, Moses T, Akamatsu S (2000) Characteristics of nuclei in Chinese freshwater cultured pearls. Gems Gemol 36:98–109

Scarratt K, Bracher P, Bracher M, Attawi A, Safar A, Saeseaw S, Homkrajae A, Sturman N (2012) Natural pearls from Australian *Pinctada maxima*. Gems Gemol 48:236–261

Schlüter J, Weitschat W (1991) Bohemian garnet today. Gems Gemol 27:168–173

Schwarz D, Pardieu V (2009) Emeralds from the Silk Road countries. A comparison with emeralds from Colombia. InColor 12(Fall/Winter):38–43

Shen A, Fritz E, DeGhionno D, McClure S (2006) Identification of dyed chrysocolla chalcedony. Gems Gemol 42:140

Smith CP, McClure SF, Eaton-Magana S, Kondo DM (2007) Pink to-red coral: a guide to determining origin of color. Gems Gemol 43:4–15

Soubra S (1999) Les jades de la Chine ancienne. Revue de Gemmologie AFG 136:27–30

Spencer LI, Dikinis SD, Keller PC, Kane RE (1988) The diamond deposits of Kalimantan, Borneo. Gems Gemol 24:67–80

Stachel T (2014) Diamonds. In: Geology of Gem deposits, vol 44. Mineralogical Association of Canada, Québec, pp 1–28

Strack E (2006) European freshwater pearls: origin, distribution and characteristics. Gems Gemol 42:105
Strack E (2008) Introduction. In: The pearl oyster. Springer, pp 1–36
Strack E, Kostov RI (2010) Emeralds, sapphires, pearls and other gemmological materials from the Preslav gold treasure (Xth century) in Bulgaria. Geochem Mineral Petrol 48:103–123
Szabo K, Koppel B, Moore MW, Young I, Tighe M, Morwood MJ (2015) The Brremangurey pearl: a 2000 year old archaeological find from the coastal Kimberley, Western Australia. Aust Archaeol 80:112–115
Thoresen L, Schmetzer K (2013) Greek, Etruscan and Roman garnets in the antiquities collection of the J. Paul Getty Museum. J Gemmol 33:201–222
Vollenweider M (1974) Le plus beau saphir antique. Revue de Gemmologie AFG 39:25
Webster R, Anderson BW (1983) Gems: their sources, description and identification, 4th edn. Butterworths, London, 1006 pp
Whittington S, Vose J, Hess C (1998) Emerald man. Archeaology 51:26
Wolfe AP, Tappert R, Muehlenbachs K, Boudreau M, McKellar RC, Basinger JF, Garrett A (2009) A new proposal concerning the botanical origin of Baltic amber. Proc R Soc B Biol Sci 276:3403–3412
Wyart J, Bariand P, Filippi J (1981) Lapis-lazuli from Sar-e-Sang, Badakhshan, Afghanistan. Gems Gemol 17:184–190

Chapter 3
Gem Analysis

Gemmology is the science of gems. It is a multidisciplinary field where competences coming from different areas are required: natural sciences (geology, mineralogy, crystallography, petrology, geochemistry, spectroscopy, material physics and biology) play an important role, sometimes along with human sciences (art, history and archaeology). The main scope of gem analysis is the identification of a gem, finding out the material of which it is made, whether it is natural or synthetic and if it has been treated/enhanced. The identification depends to some extent on whether the gems are in the rough state, polished, cut, and even on the way in which they are mounted. Sometimes gems mounted in jewellery limit the potential use of analysis and the conclusions are drawn using only limited methods. During the twentieth century quality grading of diamonds (e.g., 4Cs: carat, colour, clarity, cut) started to be part of gem analysis; there are various quality grading systems of other gems such as pearls or coloured gems but without such a global success. The geographic origin of gems is also an important part of gem analysis, as some stones are more in demand if they are from a more reputable or popular deposit; e.g., emeralds from Colombia are priced higher than those from Zambia, Russia, Brazil etc. and blue sapphires from Kashmir (Sumjam, Kundi Valley) are of higher value than sapphires from another mining area. Geographic origin determination can also give valuable clues regarding ethical concerns (e.g., identification of gems coming from areas considered by some people as "conflict" such as Emeralds from Swat Valley and rubies from Burma) as well as archaeological questions (e.g., help to trace ancient trade routes via geographic origin determination of gems found in jewellery of archaeological interest). In order to perform geographic origin determination, geological growth environment determination is required; e.g., metamorphic, metasomatic, basalt-related etc. and if they come from a primary or secondary occurrence if these were in their rough state.

Gem analysis should be non-destructive or rarely minimally (i.e., at micron level) destructive (Webster and Anderson 1983). This limits the choice of investigation methods as "too" destructive techniques are only used for research purposes.

The original version of this chapter was revised: New figures are updated with new captions and few figures are updated without changing their captions. The correction to this chapter is available at https://doi.org/10.1007/978-3-030-35449-7_6

Accessories used for various instruments, e.g., beam condenser for UV-Vis-NIR and FTIR spectrometers, are also important in order to take a proper spectrum. This requires highly skilled and trained scientists with experience on working with gem material. Instrumentation used for gem analysis could be divided in two major groups: tools for practical gemmology and advanced laboratory instruments. Practical gemmology requires basic tools which are fast, easy to transport and non-destructive; most of them are available in gemmological laboratories, sometimes also in jewellery shops. They are also used by buyers in gem markets. Laboratory instruments are more complex, requiring scientific background and specialized training. Often, such instrumentation is not specifically devoted to gems but has been designed for the use in other fields. Some instruments appear in a portable and/or mobile version (i.e., more lightweight and smaller footprint), usually with lower performance. These versions are useful to study unmovable precious gems and jewels present in collections and museums or to characterize gems directly in the field, in their deposits or outcrops. In some cases, large facilities for research scopes are also used.

3.1 Classic-Practical Gemmology/Basic Gemmological Tools

3.1.1 Observation Under Different Magnifications

Observation with unaided eye as well as under different magnification was and still is very important in gemmology. With unaided eye, the basic characteristics of a gem can be described, such as colour, pleochroism (when strong), lustre, cut and shape. Since colour is dependent on the light source, the observer should be aware that under some light the colour may appear different and lead to wrong results (e.g., pink vs pinkish red for pink sapphire vs ruby). The magnifying loupe used by gemmologists is usually 10× (corrected) and it is sometimes combined with a flashlight and darkfield (useful for the observation of unmounted transparent stones; Boehm 2002). With the loupe, internal characteristics (e.g., inclusions, colour zoning) and/or surface characteristics are observed and certain conclusions can be drawn (e.g., the presence of natural inclusions signifies that the gem is natural, differences in surface lustre may indicate a treatment or if it consists of a doublet or a triplet). In some cases (i.e., by an experienced gemmologist) other characteristics such as birefringence, hardness of the stone (abrasion of corners and scratches), doubling -e.g., tourmaline- can be observed. Loose gems are cleaned by a gem cloth and usually (in case of hard enough gems) held using tweezers (Fig. 3.1).

Gems are also observed under higher magnification using a gemmological microscope, which is a stereo-microscope with a particular setup (Fig. 3.2). The two independent optical paths (one for each eye; i.e., binocular microscope) allow a 3-dimensional observation of the gem. Magnification is commonly 10× to 60×; often combined with an auxiliary doubling lens going up to 120×. Two independent eyepieces are collecting the image coming from a wide objective. Illumination is

Fig. 3.1 Gem cloth, tweezer and 10x magnifying loupe. (Photo: Stefanos Karampelas/LFG)

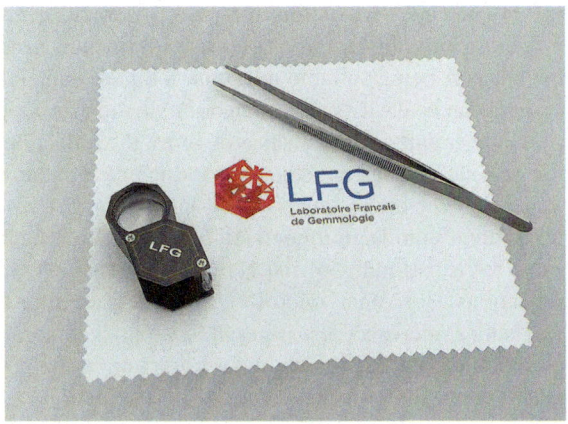

Fig. 3.2 An example of a gemmological microscope with a digital camera on the top. (Photo: Stefanos Karampelas/LFG)

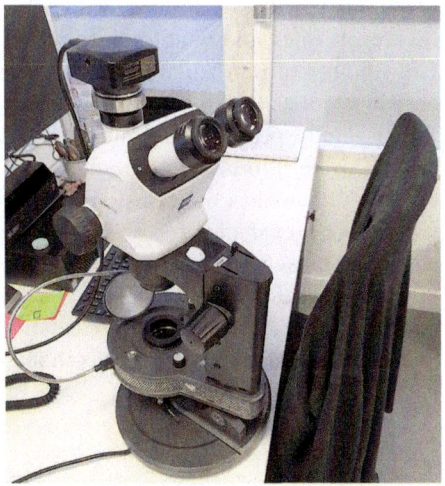

very important for the gemmological microscope as different illumination/lighting can reveal different internal and external features. The lighting system should comprise transmitted light, in both dark field and bright field. Transmitted light illumination is important for the observation of inclusions and inner structures. Sometimes when observing a gem in bright field, a diffuser is required to reveal the colour zoning present in the stone. A source for reflected light illumination is also present, necessary for opaque materials, mounted gems, study of surface characteristics etc. The source for reflected light can be mounted on a flexible arm; fibre optic (concentrated light) illuminators are also used because they allow more flexibility in the geometry and direction of lighting. An important add-on for the gemmological microscope are polarizing filters, which allow to use the microscope as a magnifying polariscope (see below for more information regarding polariscope). In the modern gemmological microscopes, a trinocular head is usually present, the third optical port being suitable to mount a digital camera (Renfro 2015). Otherwise, a camera can be easily incorporated into the microscope head. To enhance the vision of the inner part of the transparent gems, the reflection and refraction of the light on the gem surface can be reduced working in a medium with a refractive index higher than air. Some microscopes are designed to easily work with the gem embedded in transparent liquids, in immersion configuration. The best is to use a liquid with the same refractive index of the gem, but even simple water will strongly reduce the reflections. Horizontal microscopes are also used for this purpose.

Using a microscope, it is possible to obtain a large amount of information on the type of gem material being under observation; the presence of solid or fluid inclusions, clues for provenance/origin determination (e.g., diagnostic inclusions) or identification of synthetics (e.g., curved striae in corundum), detecting treatments (e.g., fillers and coatings), identifying typical simulants obtained by combining different parts as in doublets and triplets. Another application is the differentiation of mammoth from elephant ivory when observed in reflection, based on their growth (Schreger) lines, their angles are <90 and >115 respectively (Hodgkinson 2015). Growth lines of pearls like fingerprints (overlapping aragonite platelets) give valuable indications that it is a pearl (nacreous) and not an imitation. Also candling (observing the pearl in transmitted light) and observation through the drill hole can give valuable information on cultured pearl identification. In the series of books entitled: 'Photoatlas of inclusions in gemstones' by Gubelin and Koivula (1986, 2005, 2008), examples for observations made through the gemmological microscope are presented (see also Fig. 3.3). A keen understanding of inclusions and other internal characteristics is essential though when viewing the interior of a gem (Boehm 2002). An important application on mounted stones with a closed back setting is the calculation of gems' depth to assist in the calculation of its weight. This

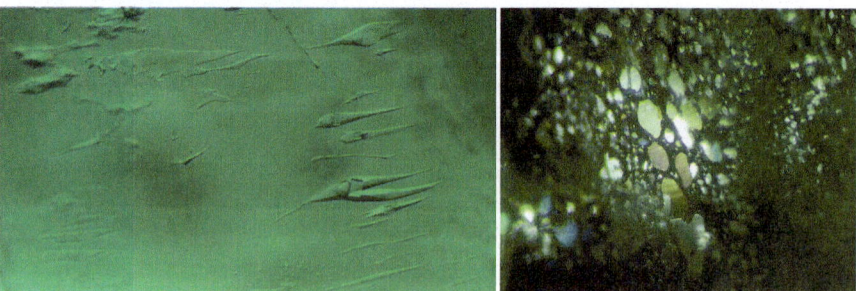

Fig. 3.3 Photomicrographs of emeralds from different locations. Multiphase inclusions with jagged outlines in a natural emerald from Colombia (left photo; field of view: 1 mm) and iridescent thin films in a natural emerald from Russia (right photo; field of view: 2.2 mm). (Photomicrographs: Ugo Hennebois/LFG)

can be done measuring the apparent maxima, by exactly focusing on their surface and on the bottom and reading the values from a vernier. The apparent maxima should be multiplied by the refractive index (see below) of the gem in order to calculate the depth.

3.1.2 Refractive Index Measurement, Optical Signs and Specific Gravity

The refractive index (R.I.) n is defined as the ratio between the speed of light in vacuum and the speed of light in the material. The measurement of the R.I. can be performed easily, with a refractometer. It is a simple instrument, based on the measurement of the critical angle at which the total reflection of a faceted gem occurs. In a classic gemmological refractometer, the gem is placed on a good optical contact on the top face of a glass prism (with R.I.: 1.96) using a drop of a high refractive index liquid. R.I. of the used liquid represents the maximum value measured with this technique. Total internal reflection due to the gem can only be measured if the gem being tested has a lower R.I. than the glass prism and the contact liquid. The most used liquid is a saturated solution of sulphur, di-idiomethane and tetraidioethylene ($n = 1.81$) or saturated solution of sulphur and di-idiomethane ($n = 1.79$); it should be used with great care, avoid skin contact, wash hands after use and use in properly ventilated conditions not to inhale the vapour. The standard measurements are obtained using a monochromatic yellow light with a wavelength of 589.3 nm or yellow emitting diode (LED) and a polarizing filter (Fig. 3.4). The value of the refractive index can be read on a scale, with a precision of the second decimal, usually the third decimal can be estimated at those refractometers generally used by gemmologists. Repeated measurements should be taken table down (table facet, when present, on the glass where the liquid is), rotating the stone (four times every 45 degrees) as well as followed by a couple of measurements on another side and reading of the shadow edges. If only one shadow at the same position is observed, the gem is isotropic. Only amorphous materials (as glass) or gems belonging to cubic crystal family (diamond, garnet, spinel) are singly refractive, meaning that their refractive index is not dependent on the direction of propagation of the light. Many gemstones are anisotropic and birefringent, showing different values of refractive index depending on the directions of propagation and polarization (the direction of the electric field transported by the light). They can be uniaxial with two refractive indexes, the refractive index along a crystallographic axis (c-axis) different from the refractive index along the other two identical crystallographic axes (tetragonal or hexagonal crystal family gems) or biaxial, with three different refractive indexes values along the three crystal axes (orthorhombic, monoclinic or triclinic crystal family gems). In both cases (uniaxial and biaxial gems) the shadow

Fig. 3.4 Refractometer with an integrated light source and liquid commonly used to measure the refractive index of a gem. (Photo: Stefanos Karampelas/LFG)

edge moves between two values on the scale and their positive or negative character can be identifed. In some cases, even heat-treated tourmalines can be identified by experienced gemmologists for example by using Kerez effect/satellite readings (Koivula et al. 1994; Sturman 2005). The difference between refractive indices give gems' birefringence. Anisotropic gems observed though the optic-axis (c-axis) could show only one line; because of this, gems should always be checked in different angles. When a gem is not faceted but has only curved surfaces such as cabochon or carved, the refractive index can be measured by spot reading (distant vision method), which is less accurate and with which usually the birefringence cannot be measured. Porous, permeable or chemically unstable gems, (e.g., chalcedony, turquoise, pearls, corals, ivory, amber) should be measured with great care as the contact liquid could affect them.

3.1 Classic-Practical Gemmology/Basic Gemmological Tools

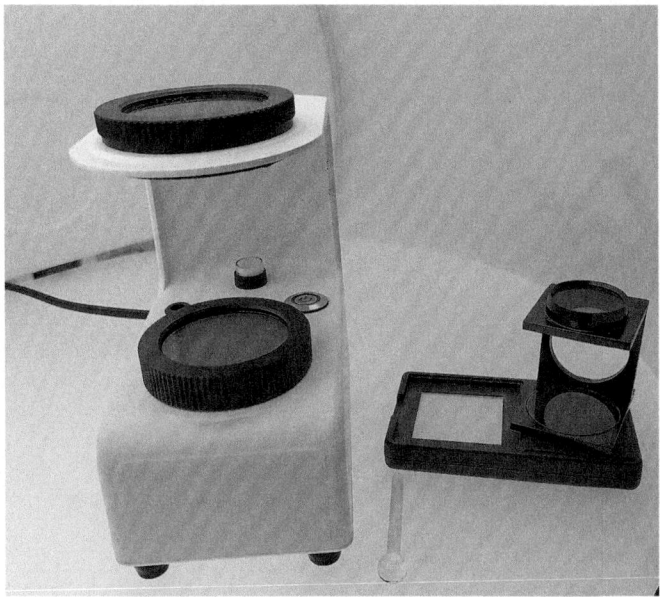

Fig. 3.5 A polariscope on the left with the rod used as conoscope. On the right a mobile polariscope on a light source. (Photo: Stefanos Karampelas/LFG)

Polariscope is another important simple instrument used to assist in the identification of transparent to some translucent gem materials. The gem is placed over a fixed polarizer, usually illuminated from the bottom, with a rotating polarizer (analyzer) on the top (Fig. 3.5). When polarizers are in crossed configuration (i.e., their axis are mutually perpendicular) the field between them appears black. Observations should be made with crossed polarizers while the gem is rotated by 360 degrees. If the gem appears always black (extinction) it is optically isotropic, either amorphous or crystallizing in the cubic system (e.g., glass, plastics, opal, diamond, garnet, spinel). If the gem blinks four times light and dark while rotating, it is optically anisotropic (uniaxial or biaxial). This should not be confused with the anomalous extinction effect (a.k.a. anomalous double refraction) which is wavy and could appear in some synthetic (verneuil) spinels, almandine garnet, some amber or other resins, some diamonds, some glasses and plastics. If the stone appears light while rotating it is a microcrystalline or cryptocrystalline aggregate (e.g., chalcedony, "jade"); repeatedly twinned gems (e.g., sapphires) and doublets could also show this pattern. Conoscope (a glass sphere on a rod) in combination with a polariscope, helps to better identify the optic character of anisotropic (uniaxial vs biaxial) stones as well as to observe the interference figures, situate crystallographic axis etc. This has various application in gemmology; e.g. identification of (natural and synthetic) quartz with a bull's eye pattern. With the use of plates made of mica, gypsum, plas-

Fig. 3.6 Dichroscope on the left and colour filter on the right. (Photo: Stefanos Karampelas/ LFG)

tic or other material (retardation plate), positive or negative character of anisotropic gems can be identified. This can further help to gem identification.

Dichroscope is another basic tool that gives valuable information on the identification of transparent to translucent gems (Fig. 3.6). It is used to observe the different colours, if present, of a possibly pleochroic gem (a gem showing different colours when observed at different angles).Those used in gemmology are made of calcite (two colours are observed at once) or two polarizing filters placed side-by-side, with their polarizing directions at 90 degree angle to each other. The gem should be observed along the different angles perpendicular between them. If the colours in all directions are the same, the gem is isotropic. If two colours (dichroic) are observed the gem is anisotropic and uniaxial and if three colours are seen the gem is anisotropic and biaxial. Conclusions should be drawn with great care as sometimes the difference between pleochroic colours can be difficult to observe and some colours are affected after particular treatments.

The weight of gems is usually expressed in carat and it is measured with a digital carat balance which can take readings to two to three decimal places. One carat (1 ct) weighs one fifth of a gram; 1 ct = 0.2 grams. The carat weight should not be confused with carat proportions in gold, which is a totally different unit. The weight for mounted gems can be estimated without unsetting them. For this the gem's dimensions and identity of the material is required to enable the user to get specific gravity of the gem material (Table 3.1). The formula used may be slightly different

Table 3.1 Weight estimation formulas

Cut style	Formula for weight calculation
Cabochon	L × W × D × S.G. × 0.0026 (0.0029 if very shallow)
Oval	L × W × D × S.G. × 0.0020 (0.0029 if very shallow)
Rectangular	L × W × D × S.G. × 0.0027

L Length, *W* Width, *D* Depth, *SG* Specific Gravity

depending on the gem's shape. Some websites on the internet as well as some articles and books (e.g., Carmona 1998a, b) can be used as reference as well.

Specific gravity (S.G.) is defined as the ratio of the weight of a substance in air to the weight of an equal volume of water. Specific gravity can only be measured on unmounted gems using the hydrostatic method with a modified digital carat balance; S.G. = A/(A–W) where A: gem's weight in air and W: gem's weight in water. Most gems have a S.G. between 2 and 4. The exceptions are the very dense zircon (4.7) and, on the light side, organic gems such as amber, with a density not far from that of water (slightly above 1). Mixture of water with salt can be used to separate plastic material (imitation) from amber. A combination of S.G. and R.I. can also help to differentiate between various garnet species.

High density liquids can also be used to measure S.G., a method useful for small unmounted gems. Starting from liquids with different specific gravities, it is possible to mix them in different proportions, in order to obtain a solution with the required specific gravity to discriminate between the desired gems. As an example, in a liquid (or mixture) with specific gravity of 2.8 g/cm^3, citrine quartz crystals (s.g. = 2.7 g/cm^3) will float while topaz crystals (s.g. = 3.5 g/cm^3) will sink. However, due to their potential toxic nature and the possible damage to many kinds of gems (porous or fissured, coloured with organic pigments, clarity enhanced, etc.) the use of high-density liquids in gemmology is limited.

3.1.3 SW- and LW-UV Fluorescence

One of the easiest and most widespread methods to test gems is fluorescence under short wave (SW-) and long wave (LW-) ultraviolet (UV) lamp excitation. The term fluorescence refers to the absorption of high-energy radiation with the consequent emission of lowest energy radiation (that means at lower frequency, or at higher wavelength). There are also other types of fluorescence, e.g., after X-ray excitation or even after excitation with visible light. In gemmology, the most used sources for the observation of fluorescence are 3 Watt lamps emitting at 254 nm (SWUV) and 365 nm (LWUV) in a dark viewing cabinet where the gem is placed at *ca.* 10 cm distance from the lamp and the viewing glass. It is strongly recommended to check

Fig. 3.7 Ruby generally displays a distinct long-wave UV fluorescence. This UV fluorescence is strongest in rubies with higher chromium vs iron ratio such as rubies from Myanmar (2 stones bottom right), and weakest in rubies with lower chromium vs iron ratio such as rubies from Thailand (marquise shaped in the back). Length of the cushion-shaped ruby on the left: 5.8 mm. (Photo: Lore Kiefert)

the fluorescence via UV protective goggles, as SWUV light is harmful to the eyes and skin (when it is of high power). It is possible to detect fluorescent fillers, to recognize fluorescence due to some treatments (e.g., the chalky blue fluorescence of synthetic or heated sapphires under SW excitation) as well as to obtain confirmation on the provenance of some gems. For example, marble hosted rubies from Myanmar, Vietnam and Afghanistan will show a more intense fluorescence than those from Thailand or East Africa as the formers contain more chromium and less iron (higher Cr/Fe ratio; see Fig. 3.7). The technique is also useful to quickly check diamonds; for instance more diamonds show a weaker fluorescence in SW than in LW and if a diamond presents more intense fluorescence in SW than in LW is most likely synthetic. Moreover, in general the reaction of diamonds is very variable so if a series of transparent stones presents similar fluorescence under SW -and LW-UV, it is very probable that these are not diamonds.

In the last years laser pointers and LEDs at different wavelengths in the UV and visible range appeared in the market. Violet sources, at about 405 nm, are the most used. Even if compact and easy to handle and used to produce fluorescence, all lasers are highly dangerous for the eye and should be only used with suitable protective goggles and without people around. Also, due to their high intensity, the results cannot be compared with standard UV lamps (usually even weakly fluorescent gems show a high fluorescence when illuminated with a laser pointer).

3.1.4 Handheld Spectroscope

To analyse the colour of a gem, a handheld visible spectroscope and a UV-Vis-NIR spectrometer can be used. The first is fairly easy to use while the second belongs more to laboratory techniques and it is explained at a later stage with more information on the origin of colour of gems. There are two types of handheld spectroscopes used in gemmology, the prism and diffraction grating spectroscopes, differing just in the element used to disperse the light (to separate the different colours) as well as the projected scale (scale using prism grating is not linear with the violet part more important and the red less important whereas using diffraction grating the scale is linear). Both of them consist of a small tube with a slit at one end and a lens at the other. A strong white light source is used in transmitted (where the stone is placed in between the light source and the spectroscope) or reflected light. The choice of light source should be done with caution as bands present in the produced light could be confused with bands due to the light absorption of the gems. When the light passes through the spectroscope without a gem, the observer should only see rainbow colours (violet, blue, green, yellow, orange and red –from around 400 to 700 nm-). When the light passes through a gem some parts of the spectrum are missing, i.e. the part that the gem is absorbing. Similar looking coloured gems can give different patterns, e.g. ruby, red spinel and garnet spectra differ from each other. Not all coloured gems produce a distinctive pattern and most synthetics present spectra similar to their natural counterparts. Some colourless gems though show diagnostic spectra (e.g., diamonds, zircon). Using this method, transparent to opaque gems can be checked even if they are mounted.

3.1.5 Other Classical Techniques

In addition to those previously shown, few other diagnostic techniques requiring simple apparatus are used in classic gemmology. Some are based on the measurement of properties of the materials such as magnetic susceptibility (paramagnetism or diamagnetism) by using simple to more complicated methods. For instance, red garnets will react to a magnet whereas red spinel and rubies will stay inert. Peridot and green tourmaline are magnetic as well. Some synthetic diamonds are also magnetic due to their impurities.

Thermal conductivity is another characteristic sometimes used for gem identification. It is used to separate diamond from its simulants (e.g. cubic zirconia a.k.a. CZ) since diamonds (natural and synthetic) are good thermal conductors, but most imitations are not (e.g., synthetic moissanite used to imitate diamond is also a good thermal conductor). Using a thermal probe is therefore considered as a quick method to draw a conclusion. Electroconductivity, with a probe, is used to separate (natural and synthetic) diamonds from synthetic moissanite as most diamonds are not electrically conductive (with exception the rare Type IIb diamonds which are conductive) while synthetic moissanite is conductive.

In the evaluation of optical properties, in particular of the colour, a series of optical filters devoted to the identification of a specific gem was developed during the years. In particular, the well-known Chelsea colour filter was designed at the end of the 1930s to separate emerald from its imitation of that time (also referred to as emerald filter; Fig. 3.6). When an emerald is viewed through the filter it appears reddish-pink to red (due to the presence of Cr), while many other green gems as well as green glass or doublet/triplets have a different reaction. However, some emeralds (e.g., from Zambia) present a different reaction as well and in parallel most synthetic emeralds cannot be identified using this filter. There are other applications of this filter, e.g., for Cr-chalcedony, Co-spinel identification, but it should only be used by experienced gemmologists with caution. Additionally, the doublets-triplets can still easily be identified even when mounted. Different filters can also be useful, e.g., Hanneman filters are used for aquamarine, ruby, tanzanite and jadeite.

Other methods, using relatively easy to find tools, were developed, e.g. "Visual Optics" system, combing different gem optic characteristics, but only well-trained gemmologists can use them. Additionally, there are some destructive tests, i.e. with a damage potential, very rarely used in gemmology such as the acid test (identification of carbonate gems, lapis lazuli etc.), hot point test (to identify amber after smelling), hardness etc.

3.2 Laboratory: Non-destructive Techniques (and Mobile Counterparts)

3.2.1 Ultraviolet-Visible-Near Infrared (UV-Vis-NIR) Spectroscopy

One of the most evident and attractive properties of a gem is the colour. Colour arises from the different absorption of light in a gem at different wavelengths (Fritsch and Rossman 1987, 1988a, b). In most cases this phenomenon is due to the presence of transition metal ions acting as chromophores (usually one or more of the following: Mn, Fe, Co, Ni, Cu, Ti, V, Cr) as part of the gem's chemical composition (idiochromatic e.g. iron in peridot) or in form of trace impurities (allochromatic e.g., chromium in emerald). The same chromophore ion can give rise to very different colours in different gems due to different crystal environment; e.g. Cr^{3+} is responsible for the green colour in emerald and of red colour in ruby. The chromophore can also give different colour due to different ion valence; e.g. Fe^{+2} give a bluish colouration to a beryl (i.e., aquamarine) and Fe^{+3} gives yellow to the same gem (i.e., heliodor). Additionally, a combination of more than one ion can influence gem's colouration; e.g., the coupling between pairs of ions as Fe^{+2}-Ti^{+4} or Fe^{+2}-Fe^{+3} as well as Fe^{+3}-Fe^{+3} plays an important role in the colouration of blue sapphires.

Transition metal ions are not the only origin of colour in gems; other causes of coloration are defects, also called colour centres (e.g., some coloured topaz and quartz), transitions between electron bands (e.g., some coloured diamonds), physical phenomena such as light interference and diffraction (e.g., opal's play of colour, labradorite) and involvement of organic pigments (e.g., some coral and pearls). For these reasons, when illuminated with white light, coloured gems will present defined absorption bands (regions of the visible spectra where they have large absorption of light). Those regions usually correspond to minima in the intensity of the transmitted and reflected light.

The analysis of the light can be performed acquiring a spectrum. The light source should be as "white" as possible, in the sense that it should contain all wavelengths of the visible light, without abrupt variations. Usually the analysis is extended to the ultra-violet (around 200 nm) and near infrared (around 1000 nm and sometimes up to 2500 nm) range, and the light source should emit down to 200 nm of wavelength. The spectra can be acquired in absorption on transparent to translucent gems, or in reflectance on opaque gems or on gems mounted with a closed back. The shape of the spectra, characterized by the positions (usually in wavelength) of spectral features such as absorption or reflectance bands, can be sometimes distinctive to identify the gem. A spectrometer in the UV-Visible range is basically composed of:

- Light sources;
- Some optics (lens, mirror) for the illumination of the sample, the light collection, and to bring the light to the detector;
- A dispersive element to separate the wavelengths (in the old spectrometers it was a glass prism, all the modern spectrometers use a diffraction grating);
- Detector(s);
- A PC to collect, show, elaborate and store the spectrum.

There are several types of spectrometers combining different sources and detectors (Fig. 3.8). Spectrometers can be of different size, weight and performances: bulky laboratory spectrometers have better sensitivity, spectral and spatial resolu-

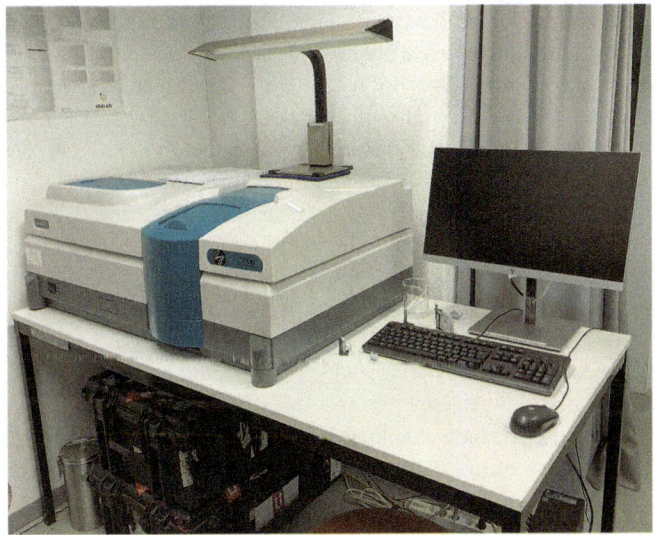

Fig. 3.8 UV-Vis-NIR spectrometer which can measure from 200 to 3300 nm. (Photo: Lore Kiefert/Courtesy: Gübelin Gem Lab)

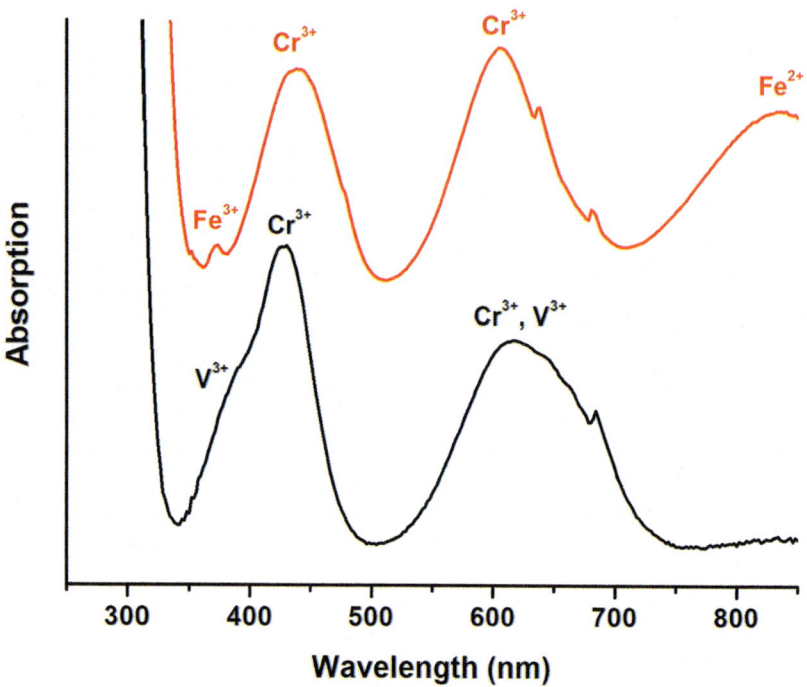

Fig. 3.9 UV-Vis-NIR spectra (not polarized) from 250 to 850 nm of an emerald from Colombia (black line) and from Zambia (red line). The latter presents characteristic absorptions linked to iron, absent in the spectrum of the first. (Courtesy: LFG)

tion and could be coupled with a microscope. Portable and handheld spectrometers are usually very light, inexpensive and easy to use in every environment but mostly with low spectral resolution. Spectral resolution is dependent of spectral bandwidth (SBW) and data interval (DI) of the spectrometer. A low scan rate (or higher acquisition times for the portable and handheld spectrometers) can give better signal/noise ratios. Very compact versions are small tubes working in direct vision without electronics for storage and processing (see again 3.1.4). Some instruments are designed for specific needs using the visible part of the spectrum, e.g., the DiamondSure, based on 415 nm absorption line present in most natural diamonds and absent in the vast majority of synthetics, is used to perform a first diamond sorting. Polarisers are frequently used to acquire spectra on some anisotropic gems. Absorption spectra, sometimes polarized, can give valuable indications on the geological environment of blue sapphires and emerald, based on Fe-related absorption bands (Fig. 3.9; see also some examples in Hanni 1994; Smith and Darenius 2009; Smith 2010; Schwarz et al. 2011; Karampelas et al. 2019b). Additionally, some colour treated gems can be identified with UV-Vis-NIR spectroscopy (e.g., green jadeite jade colour treatment). As cited above, a diffuse reflectance accessory is used to acquire spectra on opaque gems (e.g., natural and cultured pearls, corals etc.). Identification of colour treatments and sometimes host mollusc identification can be performed on pearls (Karampelas et al. 2011).

3.2.2 Raman Spectroscopy

Raman spectroscopy is a technique which is used to identify gems in a completely contactless and non-destructive way by means of the identification of their vibrational frequencies. Raman spectroscopy, and in particular micro-Raman spectroscopy, is gaining a lot of space in gemmological analysis due to its ease of use, the lack of sample preparation, the short time required for the analysis and the micrometrical spatial resolution when a microscope is used. In a Raman spectroscopy measurement, monochromatic incident light (produced by a laser) is scattered by the sample. The difference in frequency between the scattered and incident light is due to an interaction between photons and the molecular vibrations in the sample. Analysing the spectra of scattered light makes it possible to obtain the frequencies of such vibrations, which in turn depends on the mass of the atoms, resulting from the strength of their chemical bonds and on the local geometry of the crystals/crystallography or molecules. The Raman spectra, containing all different vibrational peaks of the analysed material, is a characteristic of each substance, solid (crystalline or amorphous), liquid or gas, and could be considered as a fingerprint which is of great use for its identification.

Basically, a Raman system is composed by a laser used as monochromatic light source, some optics to bring the light to the sample and to collect the scattered light, a filter to reject the light scattered without frequency change (too intense and not containing useful information), a dispersive grating spectrometer for the separation of the different wavelengths and a CCD detector. In a Raman apparatus, more than one laser is often present. The main reason is the possible presence of undesired luminescence effects, masking the Raman signal and the best way to reduce luminescence, obtaining a "clean" spectrum, is to change the excitation laser. In most cases, the use of a longer wavelength (as the 785 nm line of a laser diode or even the 1064 nm line of a Nd:YAG laser) is helpful to reduce luminescence. But in some cases, as for Cr^{3+} containing gems (emerald, ruby, spinel) the use of a green (e.g. 532 nm) or blue (e.g. 473 nm) laser allows to obtain clean spectra far from the chromium luminescence region (around 690 nm; exact position depends of the gem). Raman with lasers emitting in UV (e.g., 325 nm) can also be used, however UV lasers are in general difficult to operate. Frequently Raman spectrometers are used to obtain photoluminescence (PL) (emission) spectra (below more information on this).

At present, the most common configuration is represented by the micro-Raman spectrometer, where the focusing and collection of the light is collected through a microscope, obtaining a micrometric spatial resolution. The use of confocal optics allows to select the depth, inside a transparent material, from where the Raman signal is collected. This is very important for the analysis of inclusions in gems (which can give valuable indications on their host geology; see Bersani and Lottici 2010; Kiefert and Karampelas 2011 and references therein) and the detection of clarity enhancement (e.g., oil in emeralds; see Johnson et al. 1999; Kiefert et al. 1999). Raman spectroscopy also assists in obtaining in a very short time (seconds to minutes) the identification of a mineral species or even gem variety. In most cases, a simple comparison with free online spectral databases is sufficient for the identification (e.g., rruff.info). The separation of gems from their simulant counterparts of different chemical composition is then obtained in a quick manner. Sometimes, for gem materials like opal and amber which are considered as a bad Raman scatterer, but the use of long measurements with FT-Raman (1064 nm) can sometimes give reliable results (see examples in Kiefert and Karampelas 2011; Fritsch et al. 2012). The micrometrical spatial resolution of micro-Raman spectrometers is helpful to distinguish the different parts of a composed simulant, like a doublet or a triplet, when unmounted: the glass or quartz parts covering the slice of a precious stone are easily detected, as well as the microscopic layer of cement present in between. Raman spectra easily distinguish even polymorphs (e.g. calcite from aragonite; see some examples in Fritsch et al. 2012) but the most frequent application of Raman is the quick identification of gems when mounted and ones in which the "classical" gemmological methods described above cannot help.

Sometimes a gem is not represented by a single mineral species like, for example, corundum which is always Al_2O_3, but is a member of a mineral series or of family, as in the cases of garnet and olivine (see respectively Bersani et al. 2009; Kuebler et al. 2006), which can be determined by Raman spectroscopy. It is also used to identify synthetic as well as heat-treated spinel as they are usually disordered compared to natural untreated spinel, separation of opal-A from opal-CT as well as identification of natural turquoise and separation from imitations (Fritsch et al. 2012). Some valuable clues on emerald geology and origin as well as on the separation of natural from synthetic emeralds, based on their water and alkali elements concentration, can be also acquired (Huong et al. 2010, 2014; Bersani et al. 2014; Karampelas et al. 2019b). Raman spectroscopy is also useful for the organogenic gem identification (e.g. amber vs copal; Tay et al. 1998) as well as identification of pigments in corals and pearls (natural vs treated colour separation), and in some instances the coral genus and pearls' host mollusc identification (Karampelas et al. 2009; Fritsch et al. 2012). For Raman measurements on organic gems, the laser power used should be low in order to avoid any damage of the gem due to overheating.

In recent years many mobile Raman systems have been developed from very compact handheld systems to more sophisticated spectrometers equipped with fibre optic heads (Figs. 3.10 and 3.11). Even if the spatial and spectral resolutions are lower than those of laboratory equipment, mobile Raman spectrometers constitute a powerful solution when it is required to analyse gems in their environment, in nature or in museums or collections (Barone et al. 2014, 2016; Jehlička et al. 2017). The mobile systems are often battery operated and easy to transport and handle, with weights down to only a few kilograms, so they are very easy to transport in the field (Vandenabeele et al. 2014). The short measurement time allows to have a quick overview of a large amount of samples in a short time. The main limitation of mobile instruments is the spatial resolution, not allowing the microscopic analysis of inclusions or fissures, they commonly have one laser and relatively lower spectral resolution defined principally by the slit and diffraction grating. In order to increase the signal to noise ratio longer acquisition time and cycles are required.

3.2 Laboratory: Non-destructive Techniques (and Mobile Counterparts) 55

Fig. 3.10 Analysis on Roman jewels performed with a handheld Raman spectrometer in the museum "Laus Pompeia" (Lodi Vecchio, Italy). (Photo: Danilo Bersani. Reproduced with the permission of "Ministero per i Beni e le Attività Culturali e per il Turismo – Soprintendenza Archeologia, Belle Arti e Paesaggio per le Province di Cremona, Lodi e Mantova, Italy")

Fig. 3.11 Raman spectra performed with a portable spectrometer equipped with fiber-optic on almandine garnets present in a pegmatitic outcrop in Valchiavenna (Western Alps, Italy). (Photo: Danilo Bersani)

3.2.3 Photoluminescence (PL)

The emission of light is usually stimulated by the illumination with higher-energy (lower-wavelength) radiation; this phenomenon is a kind of photo-luminescence (PL) called fluorescence. Fluorescence can be obtained using a narrow and intense UV radiation as well as a band of visible light (usually produced by a laser) and looking for the emissions at longer wavelengths in the visible range or in the near-infrared (NIR). Emission spectra are principally used in gemmology; excitation, phosphorescence as well as time-resolved luminescence spectra on gems are rarely acquired (Gaillou et al. 2010; Fritsch et al. 2012; Hainschwang et al. 2013). This is due to the fact that most gemmological laboratories do not have a luminescence spectrometer and most spectra are acquired with a Raman spectrometer (see above chapter) with which only a PL emission spectrum can be obtained. In that case the resolution and signal can be improved similar with the Raman spectra.

PL spectroscopy has several applications on diamonds, such as the study of colour centres, treatment detection and synthetic identification. However, these are of very limited use for archaeologists. Not many studies have been conducted on the luminescence of coloured and biogenic (organogenic) gems. Chromium is the best known luminescence cause of coloured gems but also manganese, rare earth elements, uranium-based minerals as well as some organic pigments such as a certain type of porphyrin. Chromium related bands at around 685 nm are used to identify heat-treated as well as synthetic spinels and the presence of luminescence due to a kind of porphyrin is used to identify the species of some pearls (Fritsch et al. 2012; Hainschwang et al. 2013).

PL imaging, i.e. luminescence patterns observed under microscope after excitation with high energy, is applied mainly on diamonds with little examples on coloured and biogenic gems. The Diamond View™ is one example of an instrument for luminescence imaging (excitation at around 220 nm), used for the separation of synthetic vs natural diamonds with limited application for archaeologists. X-ray luminescence, i.e. luminescence after X-ray radiation, is also sometimes used in gemmology, e.g., to separate freshwater from saltwater (natural and cultured) pearls as the first fluoresce and the second remain inert (Hanni et al. 2005; Karampelas et al. 2019a). However, attention should be paid to saltwater cultured pearls with thin nacre and freshwater beads which can also fluoresce as well as on some coloured freshwater cultured and natural pearls which do not present fluorescence.

3.2.4 Fourier-Transform Infrared (FTIR) Spectrometry

Infrared absorption is an investigation technique widely used in many branches of research and materials characterization. It is based on the absorption of light in the infrared range of the electromagnetic spectrum (0.7–500 μm −700 to 500,000 nm-), and in particular in the so-called medium infrared (2.5 to 25 μm −2500 to 25,000 nm-, corresponding to wavenumbers ranging from 4000 to 400 cm^{-1}). In this range, the absorption of a photon in the mid-infrared range will cause the turning on of a vibrational mode in the molecule or in the crystal lattice. In a typical infrared absorption experiment, the light of an infrared source emitting a continuous spectrum in the full analysed range of wavelengths is transmitted through the investigated sample. The transmitted light, analysed by a spectrometer, presents a well-defined intensity decrease in correspondence of the vibrational frequencies typical of the investigated materials. They appear as negative peaks (absorption bands) in the spectrum of the transmitted light or as positive peaks when representing the spectrum in form of "absorbance". In a FTIR spectrometer, a Michelson interferometer is used to analyse the light instead of a traditional dispersive spectrometer, resulting in a better signal to noise ratio. Signal to noise ratio can be improved by acquiring more scans.

The standard transmission (absorption) measurement might require a beam condenser to make the beam pass through the investigated gem; making it sometimes possible to analyse even mounted gems, but without a closed back setting. Diffuse reflectance accessories (DRIFT) are also used in some instances as beam condensers (Hainschwang et al. 2006).

The most common application of FTIR on gems is by far the characterization of diamond types based on the amount and aggregation type of nitrogen and boron (Breeding and Shigley 2009). This can help to identify some synthetic and treated diamonds. However, most of the coloured gems present total absorption up to the region at around 2500 cm^{-1}, depending on the sample. The region in higher wavenumbers is useful as FTIR spectroscopy is sensitive to the CH stretching vibrations (between 2800 and 3200 cm^{-1}). This makes it a very good technique in detecting and identifying oils, resins and organic fillers that are used in the enhancement of gems (Johnson et al. 1999; Kiefert et al. 1999; see also Fig. 4.5).

Hydroxyl (OH) groups and water molecules produce very strong absorption when present in coloured gems, in particular in the 3000–3800 cm^{-1} region. Bands in this region are useful to identify synthetic flux grown gems (e.g., emeralds and alexandrite) as well as heated corundum (see Karampelas and Kiefert 2012 and references therein). For some identifications, such as amber, copal or other "amber-like" materials, small amounts of powdered substance (0.5–2 mg) is pressed into KBr pellets to acquire FTIR spectra (Beck 1986; Abduriyim et al. 2009 and references therein). FTIR in reflectance is also useful for the study of amber and its separation from copal (Fig. 3.12). DRIFT on smaller amounts of amber (0.1–0.2 mg) was also successfully performed (Angelini and Bellitani 2005). FTIR in specular reflectance geometry is also useful for gem identification, even those mounted with closed back settings (Martin et al. 1990; Hainschwang and Notari 2008). In recent years, compact size portable FTIR spectrometers working also in reflectance geometry have become available. They present lower resolution than research-grade spectrometers (around 4 cm^{-1}), covering a slightly shorter range but they are suitable to test gems *in situ*. FTIR spectrometers equipped with microscope, operating both in transmission and reflectance modes, can also be used, but limited examples are available in gemmology.

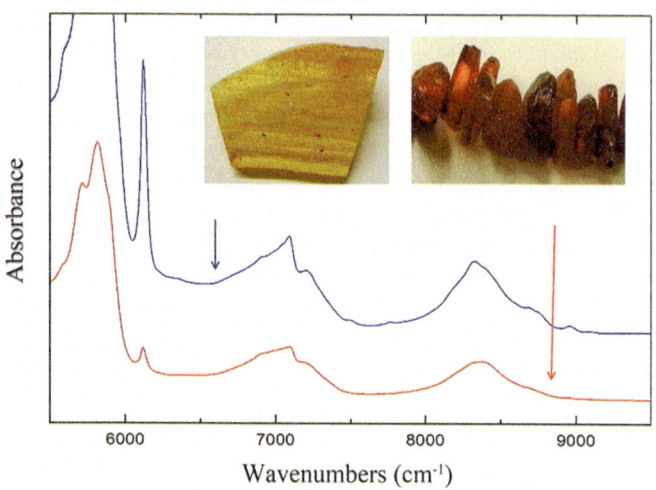

Fig. 3.12 FTIR spectrum (in the NIR region) of a Baltic amber (red line) compared with a spectrum of "younger" copal from Madagascar (blue line)

3.2.5 Energy Dispersive X-ray Fluorescence (EDXRF)

Most "elemental" techniques, used to reveal gems' chemical composition, are based on the detection of X-rays emitted by the analysed sample after some kind of excitation. The energy of the emitted X-ray depends on the difference in energy of the inner orbitals of the atoms, which are almost unaffected by their local environment or by their chemical bonds. For that reason, they can be considered characteristic of each atomic species, allowing an easy identification of the elements (but without information on how they were bounded together). To induce the emission of an X-ray, an electron vacancy is created in the first (K level) or second (L level) electronic shell of the atom. When an electron, coming from an outer shell, makes a transition to fill the vacancy, an X-ray is emitted, with an energy equal to the difference of energy of the initial and final electronic levels. To produce the vacancy, an electron in the inner shells must be removed; a suitable amount of energy can be provided by incident high energy electrons (in the Scanning Electron Microscope – SEM) or by X-rays of higher energy (in the X-Ray Fluorescence – XRF) or by accelerated protons or alpha particles or ions (in Particle-Induced X-Emission –PIXE).

In the gemmological field the most common X-Ray based elemental technique is XRF, usually with an energy dispersion detector (EDXRF). X-Ray fluorescence can be performed using laboratory apparatus, allowing a higher sensitivity (better detection limits), or portable spectrometers (Fig. 3.13) as well as handy gun-shaped instruments. The main limitation of XRF (and of all the elemental techniques based on X-Ray emission) is the very low efficiency of X-Ray emission in the light ele-

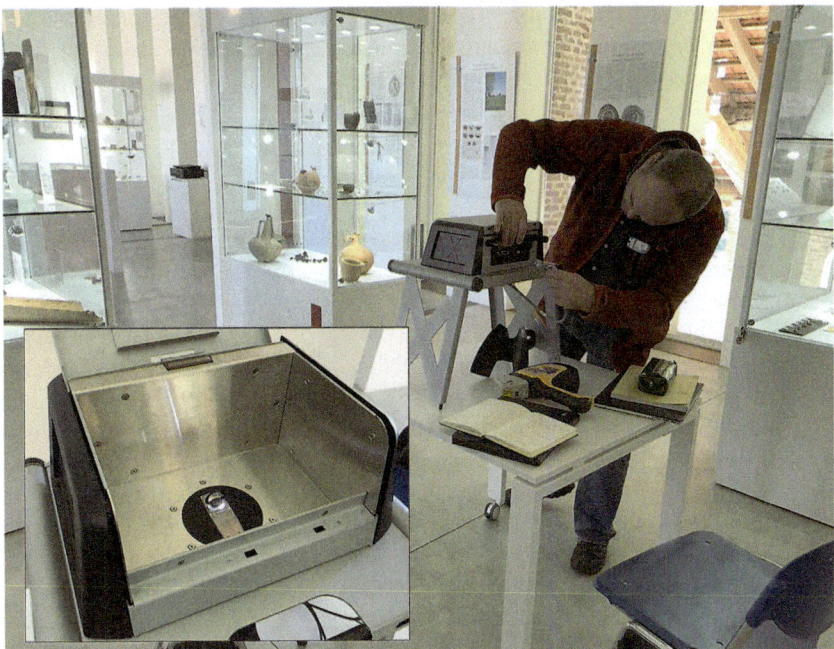

Fig. 3.13 Measurement of the elemental composition of Roman jewels by means of a portable XRF instrument in the museum "Laus Pompeia" (Lodi Vecchio, Italy). (Photo: Danilo Bersani. Reproduced with the permission of "Ministero per i Beni e le Attività Culturali e per il Turismo – Soprintendenza Archeologia, Belle Arti e Paesaggio per le Province di Cremona, Lodi e Mantova, Italy")

ments, making the detection of elements in the periodic table lighter than sodium (probably aluminium) in most cases very difficult. The instrument measures in general the bulk chemistry and should be well calibrated to produce accurate results. Attention should be paid to artefacts, such as diffraction peaks, produced by the instrument. Because of this, it is recommended to take more than one measurement on the same sample after turning it in another direction.

A classical use for XRF is the quantification of minor or trace elements to distinguish natural gems from their synthetic counterparts. In the case of rubies, the presence of tungsten, molybdenum and bismuth as well as the absence of gallium could indicate a synthetic origin, while changes in the ratio of other elements like vanadium, iron, titanium can suggest different geological environments. As an example, rubies with high iron and low vanadium content mostly come from basaltic or amphibole-related deposits while the opposite is true for rubies coming from marble-type deposits.

The different ratio between chromophore elements (V, Cr, Fe) can give valuable clues on the geological/geographic origin of emeralds, and the SrO/MnO ratio to separate saltwater (>17) from freshwater (<16) pearls. Use of portable XRF systems allow the elemental characterization (and then identification) of unmovable precious jewels preserved in museums or collections; with lower detection limits though compared with the bigger counterparts.

3.2.6 X-ray Imaging

X-ray imaging (radiography) is used to inspect the internal structures of gems, so it is important for the identification of pearls (i.e., separation of natural vs cultured pearls) and it is sometimes used for the identification of ivory. Radiography applied on gems is similar to medical radiography; it is a projection of the sample's X-ray transparency (depending on its density and composition) on an image observed on a screen or film. To create the image, a generator to produce X-rays and a detector (either digital or film) is needed and the object should be positioned in between. The detector used for gem analysis is usually of higher image resolution compared to those used for medical purposes and similar to those used by dentists. On checking a pearl's radiograph under different magnifications (micro-radiography), the organic rich parts appear darker in the radiographs and the mineral parts lighter (Fig. 3.14; see also Karampelas et al. 2017).

X-ray microcomputed tomography is also useful for gem characterization. During this measurement the sample is positioned between the x-ray source and the digital detector and rotated through 360° taking radiographs in every defined angle (e.g., every 0.5 degrees amounts to 720 projections). All acquired projections are used to reconstruct a 3D model of the sample. With the help of a dedicated software the model can be visualised. This method is useful for pearls (see some examples in Karampelas et al. 2010) and for the identification of ivory using the growth lines

Fig. 3.14 X-ray micro-radiographs of three pearls; a button shape cultured freshwater pearl without bead in the left of around 10.6 mm long with a hair-like structure in the centre which is linked to the cultivation procedure, a near round cultured saltwater pearl of around 10 mm diameter with bead in the middle made of freshwater shell of about 7 mm in diameter and a button shape natural saltwater pearl of around 8.1 mm long in the left presenting "onion ring" structures and darker centre. (X-ray micro-radiographs: Sophie Leblan/LFG)

described above whenever they cannot be observed under a microscope. Much clearer and sharper images are obtained by the use of X-ray microcomputed tomography and more details are observed which cannot be seen in the normal X-ray microradiographs.

3.2.7 *Laser Ablation – Inductively Coupled Plasma – Mass Spectrometry (LA-ICP-MS) and Laser-Induced Breakdown Spectroscopy (LIBS)*

Since around 20 years, a better detection limit for certain chemical elements was needed, as well as the measurement of light elements, in order to better study the geological origin of coloured gems (e.g., emeralds, rubies, sapphires, spinels) and to detect new treatments involving light elements (e.g., beryllium). Because of these needs, LA-ICP-MS and LIBS started to be used in gemmological laboratories. LA-ICP-MS is principally a mass spectrometer analysing the ions ablated from a tiny portion of the surface of a sample (usually not measurable even with a high precision gemmological balance) by a pulsed UV laser; the most frequently used wavelengths in gemmology are 193 and 213 nm (Fig. 3.15). The material removed by the laser

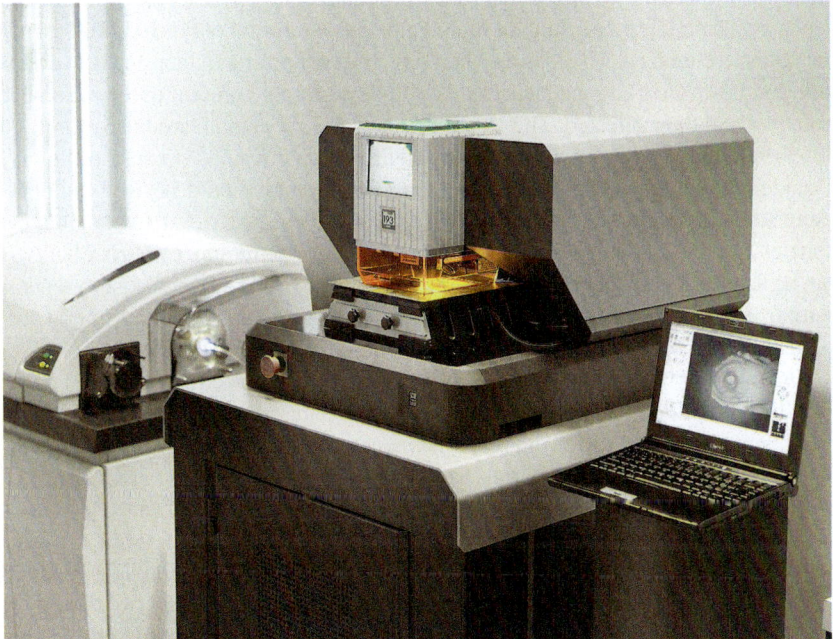

Fig. 3.15 LA-ICP-MS with a 193 nm laser used for gem analysis. (Courtesy: Gübelin Gem Lab)

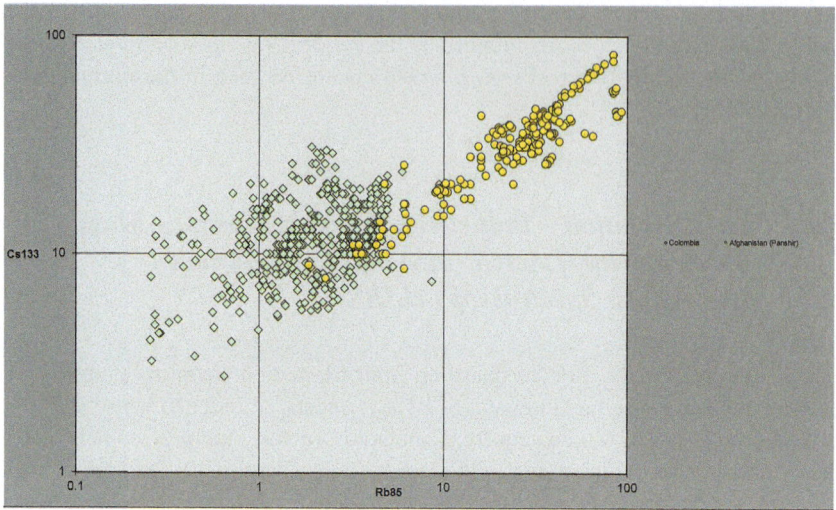

Fig. 3.16 Trace elements plot of rubidium (Rb) vs caesium (Cs) measured in ppmw with LA-ICP-MS of reference samples from Colombia (green points) and Afghanistan (yellow points). (Courtesy: Gübelin Gem Lab)

pulse is transported by a flow of gas (usually helium or argon) from the sample chamber to a plasma torch operating at high temperature (around 6000–10,000 K) where the atoms are ionized. The ions are then separated according to their mass-to-charge ratio by the electric and magnetic fields of the quadrupole present in the mass spectrometer, and then detected. The ablation craters are very small, around 3–4 holes of 30–50 microns diameter are needed to acquire reliable results; because of this it is considered as a micro-destructive method. The system should be well calibrated using adequate standards (e.g., NIST 610 for most gems, MACS-1 for calcium carbonates and doped synthetic corundum for corundum to minimize matrix effects). It can be used, among others, to find out the provenance of some gems; e.g., to separate natural emeralds from Colombia with those from Afghanistan which might look similar microscopically (Fig. 3.16; see also Karampelas et al. 2019b).

On the other hand, LIBS is a type of atomic emission spectroscopy and a pulsed laser (<200 nm lasers are used in gemmology) is focused on the surface of the analysed gem. As the short (few nanoseconds) energy pulse reaches the gem, it locally ablates a small quantity of material (from tens to hundreds of nanograms) on an area of nearly 100 micrometers of diameter; several measurement points need to be measured for accurate results. The ablated material forms a plume of very hot plasma (T > 30,000 K), composed of excited ions, electrons and fragments of molecules. After a short time (nearly 300 ns) where the continuous light emission due to the electron deceleration and electron-atom recombination dominates the spectra, typical emission lines during the electronic transitions in the excited atoms appear in the visible-UV range, allowing the identification of the elements. A sensitive spectrometer, usually a CCD camera, collects the light emitted by the plume by means of a lens system or a fibre optic. A video camera, often coupled with a microscope, allows the choice of the analysed spot.

Both methods are micro-destructive and in cut gems the analysis is therefore usually performed on the girdle. LA-ICP-MS is more difficult to operate and more expensive than a LIBS apparatus. Additionally, LIBS systems are easy to couple with Raman spectrometers and LIBS spectrometers can be produced in very compact forms: the first handheld LIBS instruments recently appeared on the market. However, LIBS is less sensitive than LA-ICP-MS and therefore most gemmological laboratories are using the latter.

3.2.8 Other Methods

In the field of material science, X-Ray analysis in a SEM (Scanning Electron Microscope) equipped with an energy dispersion (EDS) or wavelength dispersion (WDS) detector is very common, combining the high spatial resolution (less than 1 micrometer, down to tens of nanometers) with the possibility to produce chemical and morphological micro-mappings. Most SEMs require the preparation of the sample with a metallic coating to make it conductive, thus it is of limited use on gems as it is. Some instruments can also work in low vacuum, so no metallization is needed. However, the high cost of such instruments as well as the limited additional advantages compared to other instruments restrict the application on gems (Fritsch et al. 2006; Gaillou et al. 2008). PIXE (Particle-induced X-ray Emission) is not destructive and not invasive, but requires complex and costly instrumentation, as a tandem accelerator, or the use of a radioactive source to produce alpha particles. As an example, PIXE was used on a tandem accelerator on rubies and emeralds for origin determination as well as synthetic identification (Tang et al. 1988, 1989; Yu et al. 2000). Another example is the external beam setup which was used at the Aglae facility (Centre de recherche et de restauration des musées de France, Louvre, Paris) to identify the chemistry of archaeological garnets (Calligaro et al. 2002). PIXE was also used on lapis lazuli, jadeite as well as other gems (Pappalardo et al. 2005; Ruvalcaba-Sil et al. 2008; Lo Giudice et al. 2009).

Secondary Ion Mass Spectroscopy (SIMS or nano-SIMS) is a highly sensitive chemical method. It is micro-destructive, and sometimes coating is required, the sample chamber is of limited size, high cost, the instrument calibration is very sophisticated, and it needs major sample preparation (e.g., time for ultra-high vacuum). However, it gives very good results for all elements including isotopes. Oxygen isotopes were used for the geologic and geographic determination of emeralds and corundum (Giuliani et al. 2000, 2005).

Radiocarbon (^{14}C) dating was applied recently to date saltwater pearls; however above 5 mg of material -pearl- is needed (Hainschwang et al. 2010; Krzemnicki and Hajdas 2013; Zhou et al. 2017). DNA analysis for the identification of some saltwater pearls from *Pinctada* sp. as well as coral from *Corallium* sp. and ivory was performed; this method needs 5–10 mg material (Meyer et al. 2013; Saruwatari et al.

2018). X-ray diffraction of single crystals has some limited applications on gems (e.g., separation of different polymorphs). Other occasionally used techniques on gems are Transmission Electron Microscopy (TEM), Atomic Force Microscopy (AFM), Nuclear Magnetic Resonance (NMR), Electron Paramagnetic Resonance (EPR), Cathodoluminescence (CL), X-ray topography, Focused Ion Beam (FIB), Neutron Activation Analysis (NAA) analysis, Neutron radiography.

References

Abduriyim A, Kimura H, Yokoyama Y, Nakazono H, Wakatsuki M, Shimizu T, Tansho M, Ohki S (2009) Characterization of 'green amber' with infrared and nuclear magnetic resonance spectroscopy. Gems Gemol 45:158–177

Angelini I, Bellintani P (2005) Archaeological ambers from Northern Italy: an FTIR-DRIFT study of provenance by comparison with the geological amber database. Archaeometry 47:441–454

Barone G, Bersani D, Crupi V, Longo F, Longobardo U, Lottici PP, Aliatis I, Majolino D, Mazzoleni P, Raneri S, Venuti V (2014) A portable versus micro-Raman equipment comparison for gemmological purposes: the case of sapphires and their imitations. J Raman Spectrosc 45:1309–1317

Barone G, Mazzoleni P, Raneri S, Jehlička J, Vandenabeele P, Lottici PP, Lamagna G, Manenti AM, Bersani D (2016) Raman investigation on precious jewelry collections preserved in Paolo Orsi Regional Museum (Siracusa, Sicily) by using portable equipment. Appl Spectrosc 70:1420–1431

Beck CW (1986) Spectroscopic investigations of amber. Appl Spectrosc Rev 22:57–110

Bersani D, Lottici PP (2010) Applications of Raman spectroscopy to gemology. Anal Bioanal Chem 397:2631–2646

Bersani D, Andó S, Vignola P, Moltifiori G, Marino IG, Lottici PP, Diella V (2009) Micro-Raman spectroscopy as a routine tool for garnet analysis. Spectrochim Acta A 73:484–491

Bersani D, Azzi G, Lambruschi E, Barone G, Mazzoleni P, Raneri S, Longobardo U, Lottici PP (2014) Characterization of emeralds by micro-Raman spectroscopy. J Raman Spectrosc 45:1293–1300

Boehm E (2002) Portable instruments and tips on practical gemology in the field. Gems Gemol 37:14–27

Breeding CP, Shigley JE (2009) The 'type' classification system of diamonds and its importance in gemology. Gems Gemol 45:96–111

Calligaro T, Colinart S, Poirot JC, Sudres C (2002) Combined external-beam PIXE and micro-Raman characterisation of garnets used in Merovingian jewellery. Nucl Instrum Methods Phys Sect B 189:320–327

Carmona CI (1998a) Estimating weights of mounted colored gemstones. Gems Gemol 34:202–211

Carmona CI (1998b) The complete handbook for gemstone weight estimation. Germania, Los Angeles, 434 pp

Fritsch E, Rossman GR (1987) An update on color in gems. Part 1: introduction and colors caused by dispersed metal ions. Gems Gemol 23:126–139

Fritsch E, Rossman GR (1988a) An update on color in gems. Part 2: colors involving multiple atons and color centers. Gems Gemol 24:3–15

Fritsch E, Rossman GR (1988b) An update on color in gems. Part 3: colors caused by band gaps and physical phenomena. Gems Gemol 24:81–102

Fritsch E, Gaillou E, Rondeau B, Barreau A, Albertini D, Ostroumov M (2006) The nanostructure of fire opal. J Non-Cryst Solids 352:3957–3960

Fritsch E, Rondeau B, Hainschwang T, Karampelas S (2012) Raman spectroscopy applied to gemmology. In: Dubessy J, Caumon M-C, Rull F (eds) Applications of Raman spectroscopy to

References

earth sciences and cultural heritage, vol 12. European Mineralogical Union and Mineralogical Society of Great Britain & Ireland, pp 453–488

Gaillou E, Fritsch E, Aguilar-Reyes B, Rondeau B, Post J, Barreau A, Ostroumov M (2008) Common gem opal: an investigation of micro- to nano-structure. Am Mineral 93:1865–1873

Gaillou E, Wang W, Post JE, King JM, Butler JE, Collins AT, Moses TM (2010) The Wittelsbach-Graff amd Hope diamonds: not cut from the same rough. Gems Gemol 46:80–88

Giuliani G, Chaussidon M, Schubnel HJ, Piat DH, Rollion-Bard C, France-Lanord C, Giard D, de Narvaez D, Rondeau B (2000) Oxygen isotopes and emerald trade routes since antiquity. Science 287:631–633

Giuliani G, Fallick A, Garnier V, France-Lanord C, Ohnenstetter D, Schwarz D (2005) Oxygen isotope composition as a tracer for the origins of rubies and sapphires. Geology 33:249–252

Gubelin E, Koivula J (1986) Photoatlas of inclusions in gemstones, vol 1. ABC Edition, Zurich, 532 pp

Gubelin E, Koivula J (2005) Photoatlas of inclusions in gemstones, vol 2. Opinio Verlag, Basel, 830 pp

Gubelin E, Koivula J (2008) Photoatlas of inclusions in gemstones, vol 3. Opinio Verlag, Basel, 672 pp

Hainschwang T, Notari F (2008) Specular reflectance infrared spectroscopy – a review and update of a little exploited method for gem identification. J Gemmol 21:23–29

Hainschwang T, Notari F, Fritsch E, Massi L (2006) Natural, untreated diamonds showing the A, B and C infrared absorptions ("ABC diamonds"), and the H2 absorption. Diam Relat Mater 15:1555–1564

Hainschwang T, Hochstrasser T, Hajdas I, Keutschegger W (2010) A cautionary tale about a little known type of non-nacreous calcareous concretion produced by the Magilus antiquus marine snail. J Gemmol 32:15–22

Hainschwang T, Karampelas S, Fritsch E, Notari F (2013) Luminescence spectroscopy and microscopy applied to study gem materials: a case study of C Centre containing diamonds. Mineral Petrol 107:393–413

Hanni HA (1994) Origin determination for gemstones: posibilities, restrictions and reliability. J Gemol 24:139–148

Hanni HA, Kiefert L, Giese P (2005) X-ray luminescence, a valuable test in pearl identification. J Gemmol 29:325–329

Hodgkinson A (2015) Gem testing techniques. J. Thomson Colour Printers (Scotland), Glasgow, 541 pp

Huong LTT, Häger T, Hofmeister W (2010) Confocal micro-Raman spectroscopy: a powerful tool to identify natural and synthetic emerald. Gems Gemol 46:36–41

Huong LTT, Hofmeister W, Häger T, Karampelas S, Trung-Kien ND (2014) Identifying natural and synthetic emeralds by vibrational spectroscopy. Gems Gemol 50:287–292

Jehlička J, Culka A, Bersani D, Vandenabeele P (2017) Comparison of seven portable Raman spectrometers: beryl as a case study. J Raman Spectrosc 48:1289–1299

Johnson ML, Elen S, Muhlmeister S (1999) On the identification of various emerald filling substances. Gems Gemol 35:82–107

Karampelas S, Kiefert L (2012) Gemstones and minerals. In: Edwards HGM, Vandenabeele P (eds) Analytical archaeometry: selected topics. Royal Society of Chemistry Publishing, Cambridge, pp 291–317

Karampelas S, Fritsch E, Rondeau B, Andouce A, Métivier B (2009) Identification of the endangered pink-to-red Stylaster genus coral by Raman spectroscopy. Gems Gemol 45:48–52

Karampelas S, Michel J, Zheng Cui M, Schwarz JO, Enzmann F, Fritsch E, Leu L, Krzemnicki MS (2010) X-ray computed microtomography applied to pearls: methodology, advantages, and limitations. Gems Gemol 46:122–127

Karampelas S, Fritsch E, Gauthier JP, Hainschwang T (2011) UV-Vis-NIR reflectance spectroscopy of natural-color saltwater pearls from *Pinctada margaritifera*. Gems Gemol 47:31–35

Karampelas S, Alalawi A, Al-Attawi A (2017) Real-time microradiography of pearls: a comparison between the detectors used in two RTX units. Gems Gemol 53:452–456

Karampelas S, Mohamed F, Abdulla H, Almahmood F, Flamarzi L, Sangsawong S, Alalawi A (2019a) Chemical characteristics of freshwater and saltwater natural and cultured pearls from different bivalves. Minerals 9:357

Karampelas S, Al-Shaybani B, Mohamed F, Sangsawong S, Alalawi A (2019b) Emeralds from the most important occurrences: chemical and spectroscopic data. Minerals 9:561

Kiefert L, Karampelas S (2011) Use of the Raman spectrometer in gemological laboratories: review. Spectrochim Acta A Mol Biomol Spectrosc 80:119–124

Kiefert L, Hanni HA, Chalain JP, Weber W (1999) Identification of filler substances in emeralds by infrared and Raman spectroscopy. J Gemmol 26:501–520

Koivula J, Kammerling R, Fritsch E (1994) Tourmaline with atypical R.I. readings. Gems Gemol 30:198

Krzemnicki MS, Hajdas I (2013) Age determination of pearls: a new approach for pearl testing and identification. Radiocarbon 55:1801–1809

Kuebler KE, Jolliff BL, Wang A, Haskin LA (2006) Extracting olivine (Fo-Fa) compositions from Raman spectral peak positions. Geochim Cosmochim Acta 70:6201–6222

Lo Giudice A, Re A, Calusi S, Giuntini L, Massi M, Olivero P, Pratesi G, Albonico M, Conz E (2009) Multitechnique characterization of lapis lazuli for provenance study. Anal Bioanal Chem 395:2211–2217

Martin F, Merigoux H, Zecchini P (1990) Reflectance infrared spectroscopy in gemology. Gems Gemol 25:226–231

Meyer JB, Cartier LE, Pinto-Figueroa EA, Krzemnicki MS, Hanni HA, McDonald BA (2013) DNA fingerprinting of pearls to determine their origins. PLoS One 8:1–11

Pappalardo L, Karydas AG, Kotzamani N, Pappalardo G, Romano FP, Zarkadas C (2005) Complementary use of PIXE-alpha and XRF portable systems for the non- destructive and in situ characterization of gemstones in museums. Nucl Inst Methods Phys Res B 239:114–121

Renfro N (2015) Digital photomicrography for gemologists. Gems Gemol 51:144–159

Ruvalcaba-Sil JL, Manzanilla L, Melgar E, Santa Cruz RL (2008) PIXE and ionoluminescence for Mesoamerican jadeite characterization. X-Ray Spectrom 37:96–99

Saruwatari K, Suzuki M, Zhou C, Kessrapong P, Sturman N (2018) DNA techniques applied to the identification of Pinctada fucata pearls from Uwajima, Ehime Perfecture, Japan. Gems Gemol 54:40–50

Schwarz D, Mendes JC, Klemm L, Lopes PHS (2011) Emeralds from South America – Brazil and Colombia. InColor 16:36–46

Smith CP (2010) Inside sapphires. Rapaport Diamond Rep 33:123–132

Smith CP, Darenius EQ (2009) Inside emeralds. Rapaport Diamond Rep 32:139–149

Sturman BD (2005) Use of the polarizing filter on the reftactometer. J Gemmol 29:341–348

Tang SM, Tang SH, Tay TS, Retty AT (1988) Analysis of Burmese and Thai rubies by PIXE. Appl Spectrosc 42:44–48

Tang SM, Tang SH, Mok KF, Retty AT, Tay TS (1989) A study of natural and synthetic rubies by PIXE. Appl Spectrosc 43:219–223

Tay TS, Shen ZX, See SL (1998) On the identification of amber and its imitations using Raman spectroscopy: preliminary results. Aust Gemmol 20:114–123

Vandenabeele P, Edwards HGM, Jehlicka J (2014) The role of mobile instrumentation in novel applications of Raman spectroscopy: archaeometry, geosciences, and forensics. Chem Soc Rev 43:2628–2649

Webster R, Anderson BW (1983) Gems: their sources, description and identification, 4th edn. Butterworths, London, 1006 pp

Yu KN, Tang SM, Tay TS (2000) PIXE studies of emeralds. X-Ray Spectrom 29:267–278

Zhou C, Hodgins G, Lange T, Saruwatari K, Sturman N, Kiefert L, Schollenbruch K (2017) Saltwater pearls from the pre- to early Columbian era: a gemological and radiocarbon dating study. Gems Gemol 53:286–295

Chapter 4
Gem Treatments, Synthetics and Imitations

When examining a piece of antique jewellery or object for its authenticity, it is not only important to know about the types of gems that existed at the time, but also about the various types of treatments, imitations and synthetic gems. This helps to determine if the piece is authentic or if it has been tampered with, i.e., if more valuable gems in the piece have been replaced with less valuable ones or imitations. The current chapter discusses some of the oldest gem treatment methods, in addition to some of the newer ones. Imitations have been used for a very long time and are usually a lower value material of the same colour of the gem it is supposed to imitate. Synthetic gemstones appeared at the beginning of the twentieth century and therefore should not be present in ancient jewellery and ornamental objects. Both, imitations and synthetics, are also covered in this chapter.

Treatments involve all the processes used to modify the appearance of gem materials in order to make them more attractive and desirable. Imitations appeared simultaneously to gems from the very early times. They were materials looking similar to the more valuable gem material but having a different chemical composition and crystal structure (e.g., blue glass and sapphire). Synthetic gems are laboratory grown materials with essentially the same chemical composition (differences between natural and synthetic gems are at impurities level) and crystal structure of their natural counterparts.

4.1 Treatments

4.1.1 A Short History of Gem Treatments

The earliest known prime source about gem treatments and imitations was probably by C. Plinius Secundus († 79 CE). About 200 BCE, Bolos of Mendas, an Egyptian chemist, compiled a volume of papyrus rolls entitled "Baphika" ("Dyeing"), which contained various techniques of treating gems by colouring. Plinius consulted many

---

The original version of this chapter was revised: New figures are updated with new captions and few figures are updated without changing their captions. The correction to this chapter is available at https://doi.org/10.1007/978-3-030-35449-7_6

older books, some thought to reach as far back as 350 BCE (Nassau 1984). Another important compilation on recipes for gem treatments was the Papyrus Graecus Holmiensis, also known as the Stockholm Papyrus, which dates back to about 400 CE, but it was thought to have been a copy of an earlier version that Plinius may have had access to (Nassau 1984 and references therein).

According to Nassau (1984) and the references he had available, the first known treatments go back to at least Minoan times (2000–1600 BCE) and implied the use of shiny metal foils to make stones appear more brilliant, or to modify their colour. Pliny describes the use of foil to make opaque gems translucent, silver foil to back "sard" (carnelian), and brass foil to back "hyacinthus" (probably sapphire) and "chrysolithus" (probably topaz). Another way to intensify or alter the colour was the use of a coating, usually paint or varnish. Mostly this was applied to the back of a gemstone at the time. It is a process that has been refined over the centuries and is still used today. Later on, there are also reports of applying pigments to the girdle of a gem to alter its colour (Bauer 1896).

Other methods mentioned in these two oldest references of gem treatments are the "softening" of gemstones, principally diamonds, by various means probably in order to apply further treatments afterwards. Pliny mentions the use of vinegar, the Stockholm Papyrus contains a recipe using goat's blood. Various oils are described to treat gems. While Pliny only reports oil, the Stockholm Papyrus also describes the use of other oily substances such as balsam sap, Canada balsam, cedar oil, liquid pitch, resin, and wax. Oiling with coloured oil is also mentioned. The Stockholm Papyrus describes several processes where a stone is "opened" or "loosened" (cracked). However, there is no reference as to how far back in time such processes have been carried out. Opening gems could involve soaking them with or without heat in certain liquids such as blood or milk of various animals, urine, bile, garlic, honey and other substances. Usually such processes were performed to make it easier for introducing a dye into the stones at a later stage. The sugar-acid treatment, which blackens porous gems (sometimes after being "opened"), was reported by Pliny, and is still in use to date on agates and opals. In this document there are around 10 recipes involving pearls; two of them require feeding the pearls to chickens (i.e., probably bleaching them using their gastric acids).

In the Middle Ages, tinting coloured gemstones such as emerald, ruby, and sapphire, was strictly forbidden by law in Italy. However, the use of foils and other backings was permitted, as well as the tinting of diamonds (Nassau 1984 and references therein). The Stockholm Papyrus mentions findings on how a treated crystal could appear like ruby and in the work of Teifaschi (or spelled Tifaschi) from about 1240 CE on how corundum (ruby) in Sri Lanka can be treated by fire (Fig. 4.1).

However, heating gained more and more attention in the middle ages. Several scholars mentioned in their books the heat treatment that was used to make sapphire colourless. They describe crucibles filled with gold or iron with the sapphire embedded in it and heated for 24 h. Agricola (1556) does not mention the heat treatment of gems but shows images of various furnaces that were undoubtedly also used for heat treating gems. Porta (1658) describes not only the methods being used for heat treatment that turns blue sapphire colourless, but also a method that renders one side

4.1 Treatments

Fig. 4.1 The blow-pipe method to heat corundum has already been reported in the middle ages and is still used today. The sapphire crystals are embedded in a clay pellet and rest on coal. The heat treater blows into the coal for hours to days to achieve the desired result, while constantly feeding coal into the ember. (Photo: Lore Kiefert)

blue and the other colourless by covering one side with clay, heating in oxidizing and reducing conditions at the same time. Pearls were also treated in several ways, either by polishing them to enhance their shape (e.g., for "blister" pearls -pearls partially attached to the shell-), or by filling them with metal in order to increase their weight.

In the nineteenth century, more scholarly works were published describing gem treatments. The most relevant work to date was published by Bauer (1896), in which he describes all previously mentioned treatments in more detail since by then most minerals had their names, and he could describe the methods for each mineral.

The beginning of the twentieth century brought several new discoveries which were also used on gems, the most significant being X-rays, discovered by W.K. Röntgen in 1895, radioactivity in uranium by H. Becquerel a year later, and gamma rays by P.-U. Villard in 1900. In the subsequent 10 years, it was detected that colourless sapphire could turn yellow, fluorite became violet, pink topaz could fade after irradiation. The effect of exposure to light and heat together with irradiation was also described at that time, and it was discovered then that irradiation-coloured diamond did not fade (Nassau 1984 and references therein).

In the following decades, all processes previously mentioned were refined and improved, especially the heat treatment of gems and irradiation processes. With better furnaces, the possibility to control not only the temperature but also the environment, as well as the combination of irradiation and heat, the means to change gems seemed vast.

The 1980's into the 2000's were marked by new as well as improved methods for gem treatments. Traditional oils were replaced with polymer resins to fill mainly the fissures of emeralds, but also of other gems, as well as stabilizing turquoise, jadeite "jade", corals and other porous materials. Linde and Union Carbide developed a process to diffuse the surface of a sapphire with titanium to improve the colour. The process was patented by Union Carbide in 1975, and the first titanium-diffused sapphires entered the market in the late 1970's (Hughes and Emmett 2005; Hughes et al. 2017). During the same period of time, diamonds started to be treated under high pressure and high temperature in order to improve their colour.

Since then, several new methods were developed. Emeralds are filled not only putting them in oil or resin but applying additional vacuum (to get the air out of the fissures) and pressure (to enable the filler to penetrate deeper into the stone). The heating of corundum is now being carried out in controlled furnaces with the possibility to add other elements such as beryllium or lithium, which does not only improve the colour, but can change it completely. Lead glass filler with refractive indices very close to corundum is used to hide fissures and turn a piece of ruby of low gem quality (e.g., translucent with numerous inclusions) at best in a transparent expensive looking ruby. And with the introduction of nano-technology, gems can now be coated with ultrathin, virtually undetectable, layers. Table 4.1 lists the most common treatments on the gems covered in this book.

4.1.2 Foiling, Coating, Oiling and Dyeing

These four methods belong to the oldest treatment methods. Foiling goes back to Minoan times and involves the use of shiny metals to back transparent gems or glass (paste). By the early nineteenth century the art of foiling paste and gemstones was well advanced and had shined in many nations of the world. This method was mostly applied to gems in closed-back settings and was used extensively over the centuries until recent times, when a greater abundance of coloured gemstones made the method more or less unnecessary. Foiling is usually visible under the microscope in older jewellery because over the centuries, the foil gets wrinkled or loses its colour so that a patchy layer is visible behind the actual gemstone.

The coating of gems usually meant using a different colour lacquer or varnish. This has mostly been applied on the back of the gemstone or the girdle. The method was applied as far back as the Greek and Roman times, but more to beautify the gems, not to deceive. Painting or coating gems is still done to date, while foiling gems is not practiced anymore. This is mainly due to the fact that metallic looking coatings do exist, which can be directly applied to the gem. In the microscope, such coatings can be easily recognized because they wear off at facet edges and are easily scratched (Fig. 4.2). Especially in transmitted light, it is made visible as thin lines that differ in colour and lustre from the surrounding surface. First mentions of recent coatings by a sputtering process occurred in 1950 by Gubelin. Nowadays, they are

4.1 Treatments

Table 4.1 Most common gemst treatments

Gem	Treatment
Amber	Heating, autoclave, pressing
Amethyst	Irradiation and/or heating
Aquamarine	Heating
Colored Beryls	Heating, irradiation
Chalcedony	Dyeing, impregnating
Chrysoberyl	Irradiation
Citrine	Heating
Coral	Impregnating, dyeing
Diamond	Irradiation, heating (HPHT), laser drilling
Emerald	Filling (oil, resin)
Garnet	Heating
Ivory and bone	Impregnating
Jade	Impregnating, dyeing
Lapis Lazuli	Impregnating, pressing
Opal	Smoke treatment, sugar-acid treatment, impregnating
Pearl	Dyeing, heating, bleaching, filling, polishing
Peridot	N/A
Quartz	Irradiating, dyeing
Ruby	Heating, oiling, filling
Sapphire	Heating, oiling, filling
Spinel	Heating, oiling
Tanzanite	Heating
Topaz	Irradiating, heating
Tourmaline	Irradiating, heating
Turquoise	Impregnating, dyeing, pressing
Zircon	Heating, irradiating

Fig. 4.2 Colourless topaz with a red coating. The coating wears first at the facet edges and can also easily be scratched off, revealing the original colour. Width: 6.1 mm. (Photo: Lore Kiefert)

Fig. 4.3 Modern oiling apparatus for emeralds in order to effectively remove any air out of the fissures with vacuum before filling them with oil (left photo). Emerald before oiling, the fissures are readily visible (central photo). The same stone after oiling, the fissures nearly disappeared (right photo). Length of the stone: 16 mm. (Photos: Gübelin Gem Lab)

slightly more durable than previous coatings due to the combination with heat, but even some of these, especially coating of topaz, quartz or tanzanite are still prone to wear and scratching.

Mostly, coatings to enhance colour are applied to the back of a gemstone. This makes it virtually impossible to use another method than microscopy when the stone is set. If the gemstone is loose, it may be possible to detect a coating with EDXRF by not only measuring the table but also the pavilion. Often, a different chemical element responsible for the resulting colour is detectable in addition to the chemical composition of the gem. Other newer coatings are used to make a gem more durable, and a very thin layer of colourless diamond-like carbon is applied to the gemstone. Pearls are sometimes coated with an extremely thin layer of nanoparticles that give an iridescence similar to the sought-after overtones in high-quality pearls. Such coatings are nearly impossible to detect, even with instruments such as a Raman spectrometer or EDXRF. For historic items, the latter coatings should not play a role, while the coatings with paint or varnish should be easily detectable.

Applying paint to parts of diamonds were already mentioned in the Papyrus Holmiensis, but in the middle ages, it appeared to be a common practice. The method was used to change the colour of diamond, but also to "whiten" it by applying blue colour around the girdle. Other than that, diamonds remained untreated until the discovery of radioactivity at the turn of the nineteenth to the twentieth century (see 4.1.4).

The oiling of gems is also a method that goes back to BCE times and is widely mentioned in books like Pliny. Basically, all gems with porous surfaces or fissures can be oiled in order to make their colour deeper and to hide otherwise reflective fissures. Oils are not very durable and dry out, seep out, or alter over time. Oil is also susceptible to solvents and can therefore be removed easily. This is often done if an oil alters and appears brown or whitish. The gem is then refilled with new oil. Therefore, if colourless oil is detected in a historic gem, it is probably a treatment that has been reapplied afterwards. Nowadays, special apparatus is used for oiling emeralds and other gems (Fig. 4.3).

4.1 Treatments

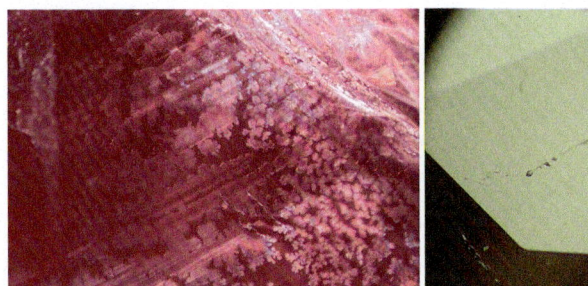

Fig. 4.4 The oil in this fissure of a ruby is partially dried out, leaving behind a patchy dendritic pattern of filled areas (darker) and reflective areas. Field of view: 0.5 mm (left photo). When heated in a microscope well or with a hot needle, droplets come out to the surface of a fissure if it is filled with oil or resin. Field of view: 1.2 mm (right photo). (Photos: Klaus Schollenbruch, GGL)

Since the late 1980's, oil has been replaced with more durable artificial resins (polymers). These resins are used in coloured gems such as emeralds instead of oils because of their higher refractive index which hides fissures more effectively, and in porous materials such as turquoise or jade in order to stabilize them (impregnation) at the same time as they enhance the colour. While such materials should not be found in historic gems, they may be used by conservators to protect the surface of an object instead of the previously used varnishes. Therefore, care must be taken to determine if it is a real treatment or only a means to conserve the gems. The use of oil or resin can often be detected in the microscope by checking carefully the surface of a gem in the microscope and look for open fissures or pores. If reflective open fissures are observed in transparent gems such as emeralds, they usually are not filled. If a fissure is filled, it is not reflective, but may show some material in it that is partially dried out (Fig. 4.4 left), showing a dendritic or patchy pattern, or in the case of a modern resin, showing subtle orange, blue or pink flashes when tilted in various directions. Sometimes when a filled gem is heated in a microscope at very low temperature, filler droplets may come to the surface (Fig. 4.4 right).

Porous gems such as turquoise or lower quality jadeite "jade" have a dull surface and light colour. By soaking them in oil or resin (i.e., impregnation), the pores fill up with the substance thus making the colour more intense. In some cases, the entire gem material may turn more transparent as well. Such treatments are much harder to detect in the microscope than filled fissures, and often can only be determined with the use of advanced instrumentation. Since oils as well as resins are organic substances, vibrational spectroscopy (such as FTIR and Raman spectroscopy) are the best methods to determine their nature. With FTIR spectroscopy, a distinction between an oil and a resin is possible in fissure-filled gems as well as in stabilized gems (Fig. 4.5). If the result of the FTIR analysis is not clear or if the object prohibits accessibility of the infrared ray, Raman spectrometry can be used, allowing larger objects as well as mounted gems to be tested.

A dye can be applied to a gem by using various methods, reaching from the above described coating to introducing dye into the gem alone or with an oil or resin. For this process, the same prerequisites of the gems - either fissures or porosity - must

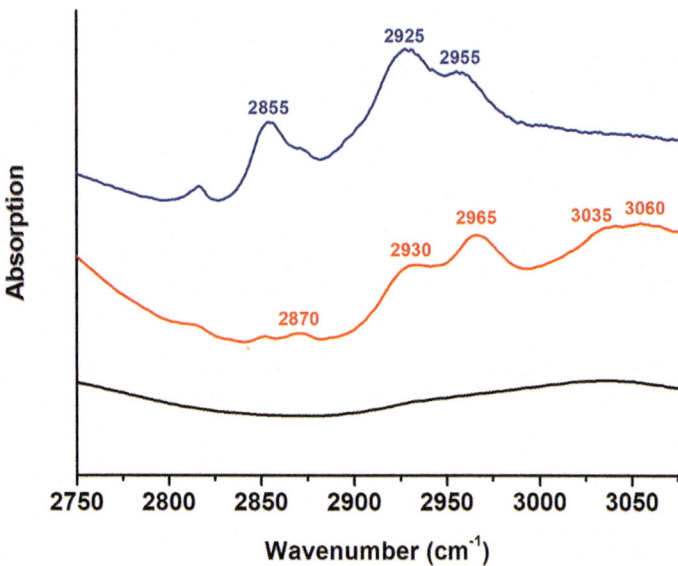

Fig. 4.5 FTIR spectra of the filling of emeralds (and other gems) with oil-type filler substance (blue line), with resin/polymer-type filler substance (red line) and of a natural emerald without any filling (black line). (Courtesy: LFG)

be given. While it is quite easy to determine dye in fissured gems, it may be much more difficult to detect a dye in porous materials or in pearls. Pearls, especially Chinese cultured freshwater pearls without bead, but also saltwater pearls with a bead, are frequently dyed. Usually the concentration of dye can be observed around the drill hole or in grooves. Silver nitrate was also used in the past to dye cultured pearls. This is absorbed by the organic matter of the treated sample and can be visible under X-ray radiography as it appears lighter due to the relatively high atomic number of the silver which blocks the X-rays more effectively than the pearl itself.

Light coloured or colourless gems which are fractured are often filled with a dye to intensify their colour. Occasionally, the colour concentration can already be seen with the naked eye, but more often such a dye only becomes obvious under magnification. Dyes can also be detected with FTIR or Raman spectroscopy if they are organic, by EDXRF if inorganic (and different from the sample's chemistry) and by UV-Vis-NIR as they have different absorption bands than their host gem. If there are no fissures with colour concentrations, it is harder to determine if a gem has been dyed. The most commonly dyed material, going back to Roman times, is chalcedony and agate. Agate in particular can be dyed in many colours, most of which are not appearing naturally.

Chalcedony is also dyed black to imitate black onyx. This sugar-acid process is another ancient treatment described by Pliny, where an acidic honey, which appears to have the combined properties of the sugar solution and the acid in more modern processes, is used to open pores, get rid of the iron oxides, and blacken the gem upon heating. As the result is carbon as the dyeing agent, it is not possible to distinguish natural black chalcedony (black onyx), also coloured by carbon, from dyed black chalcedony. Sugar-acid treatment is also used to dye the matrix of porous opal in order to highlight the colour reflexes against a dark background. However, since opal fields producing such material (Andamooka, Australia) were only discovered in the 1930's, sugar-acid treatment of opal should not be relevant for historic pieces.

Dye has been used since ancient times for other gem materials such as coral, turquoise, lapis lazuli, quartz, emerald, and "jade" to intensify their colour. The most important one here is jadeite "jade", because the deep green or violet varieties are the most sought after and can reach prices similar to rubies and sapphires. If the nature of the green colour cannot be determined microscopically, UV-Vis spectroscopy helps in most cases.

4.1.3 Heat Treatment of Gems

The heat treatment of gems has been described in early publications such as the Papyrus Graecus Holmiensis, or the work of Plinius. Although their recipes usually involve heating gems in various liquids or substances, temperatures were not as high as in modern heating processes, so the detection of such processes might not be possible even to date. Due to the vague description of the type of gems, it is also not possible to know exactly which gems were heated back then. Some ancient quartzes (citrine produced from amethyst heating of first century CE) are, however, mentioned. In the middle ages, the heating of corundum was mostly described in the literature. Abu Raihan al-Beruni mentions the heat treatment of ruby to remove the dark coloration already in his book "The Book Most Comprehensive in Knowledge on Precious Stones", which was written between 1040 and 1048 CE. Teifaschi (or Tifaschi) describes the heating of corundum in Sri Lanka in 1240, followed by other accounts in the sixteenth and then again in the nineteenth century (Huda 1998; Hughes et al. 2017 and references therein).

Even today, many of the above-mentioned heat treatments are sometimes difficult to detect under magnification and/or with modern instrumentation. Especially the heating of beryl, quartz, spodumene and kunzite, topaz, tourmaline, zircon and zoisite is sometimes carried out below temperatures where visible or structural alterations occur. Nassau (1984) gives a detailed listing of these heated gems and their effect upon heating. It was already recorded by the ancient Greeks that yellow topaz decolourized when strongly heated. However, since citrine has similar properties to topaz, and back then several gems were called topaz (see again Chap. 2), it was not clear which gemstone they were recording (Webster and Anderson 1983). Interesting in a historical context are the topaz deposits that were found in 1735 in Ouro Preto yielding intense yellow to orange and reddish-orange topaz. Soon after, in 1750, a Paris jeweller by the name Dumelle presented a method where upon heating to 500 °C the topaz became pink. Such pink topaz was very popular at the end of the eighteenth and the beginning of the nineteenth century (Misiorowski 2000). Previous to this find, pink topaz was unknown, and the major supplier was Germany, which produced yellow topaz (Nassau 1985). Another, but small, deposit of naturally pink topaz was then found in the mid-1800s in Russia, and more recently in

Pakistan (Gubelin et al. 1986). Therefore, it can be assumed that pink topaz found in Georgian jewellery most likely came from Brazil and was heated. Again, up-to-date, heat treatment of pink topaz cannot be identified by any non-destructive method.

Spinel, garnet and some other gems were traditionally not heated. However, in recent years, these gems are also undergoing heat treatment at relatively low temperatures (usually <700 °C) in order to eliminate a brownish tinge. Demantoid garnet (green coloured andradite garnet), is the only variety of the many colour varieties of garnet which might be heated. Such a heat treatment is only visible in the microscope if the fibrous inclusions have been altered or tension fissures have developed around them. Demantoid garnet only plays a role in jewellery pieces from approximately 1830 on, when the Russian deposit was found. Other demantoid deposits were discovered much later.

Spinel has been used in jewellery as long as ruby and has often been confused with this gem. Spinel occurs in many colours, however, the most sought after is the rich red colour of spinel from Burma, where it occurs in the same deposits as rubies. Some of the red spinel has a slight brownish tinge which can be reduced when the stone is heated. Usually, there is no change in inclusion scenery, however, heat does have an influence on the crystal lattice of spinel, which can be made visible with Raman and Photoluminescence spectroscopy.

Amber is an organic gem that has been valued highly in ancient times. Back then the possession of amber was considered more prestigious than wearing jewellery. Plinius gives a detailed description of the origin of amber, but also describes a clarification process where impure amber is boiled in the fat of a suckling pig (Konig and Hopp 1994). Later descriptions of amber clarification involve mostly linseed oil. Linseed oil has also been used in reconstructing amber from small pieces (Nassau 1984 and references therein). Nassau lists the following effects on amber upon heat alone: Darkening, clarification, and the development of "sun-spangle" cracks. Heat plus pressure is used for amber heating and sometimes reconstruction. In 2008, an attractive green amber appeared on the market. The colour resulted from a heat treatment in an autoclave (Abduriyim et al. 2009 and references therein), where heat together with considerable pressure was applied, but only 5 years later, natural green amber was discovered in Ethiopia (see Fig. 4.6; Kiefert 2015).

Heat has only been applied recently, after 2000, on pearls and sometimes involves heat together with dye or bleaching. The process is solely applied to South Sea (cultivated in *P. maxima* bivalve) or Tahitian (cultivated in *P. margaritifera* bivalve) cultured pearls (Elen 2001; Wang et al. 2006b). However, it can be assumed that similar processes may also be applied to Chinese Freshwater cultured pearls. Detection of such a treatment is mainly based on spectroscopy and comparison with natural coloured pearls. For historical pearls, this treatment may be neglectable because before the 1940s cultured pearls did not exist, and such treatments were not known. Only treatments to whiten/bleach natural pearls were possibly done before that time.

Up to the early 1900s, heat treatment of corundum was performed at temperatures <1200 °C and most signs of heat treatment, such as partly dissolved rutile needles or altered fingerprints are only visible in the microscope at higher temperatures (Themelis 1992). Some heated corundum shows a chalky fluorescence in

Fig. 4.6 Left: Baltic amber. Center and 5 rough pieces on top: Untreated green amber from Ethiopia. The largest piece is 6 cm wide. Bottom right: 2 pieces of Burmese amber. (Photo: Lore Kiefert)

short-wave UV light. At higher temperatures like they were used afterwards, these alterations were readily visible in the microscope. More modern heating methods with controlled furnaces make heat treatment less detectable if long heating times and different environments are used, and often analytical instruments such as FTIR on the samples or Raman spectrometers on inclusions are needed for the confirmation of a heat treatment. The heat treatment of corundum may alter the structure of inclusions in corundum from certain origins, such as boehmite and diaspore (AlOOH), which are often included in ruby and sapphire. The alteration of boehmite and diaspore is detectable in the FTIR spectrum because these minerals are transformed to Al_2O_3 at relatively low temperatures (<600 °C). So the presence of boehmite or diaspore related FTIR bands are indicative that the gemstone was not heated. Heating of corundum may also alter the appearance of crystal inclusions (Fig. 4.7 right photo). At high temperatures, inclusions usually melt and develop a fingerprint or glassy halo around them due to an expansion effect of the melted material. At lower temperatures, only subtle changes may occur. One indicator for heat treatment of corundum are zircon inclusions from geologically old deposits such as the ones formed during the East African orogeny (around 650–500 Ma). Depending on their state of metamictization, which increases with geological age due to the radioactive decay of uranium traces in zircon, Raman spectra of zircons show broad major peaks. During the heating process, the "geologic clock" is reset, resulting in narrower major peaks (Wang et al. 2006a). This detection method can only be used for corundum with ages of more than 500 Million years.

Modern heat treatment of corundum includes the addition of fluxes, various gases, beryllium and other light elements, or glasses with low melting point and a high refractive index such as Pb-glass. Although such treatments should not be encountered in historic gems, it is still important to know about them. Flux-healed

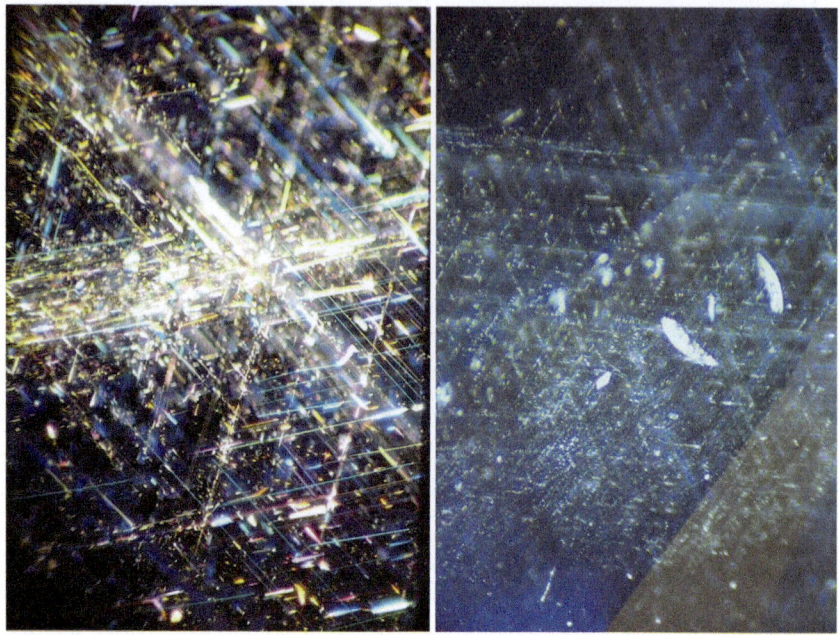

Fig. 4.7 Left: rutile needles in a sapphire from Burma. Field of view: 1.8 mm. Right: When sapphires (and rubies) with rutile needles are heated, they dissolve leaving behind a pattern of particles following the original outline of the needles. Field of view: 1.8 mm. (Photo left: Lore Kiefert, Photo right: GGL)

rubies entered the world market in the 1980's (Hanni 1992; Themelis 2004). The flux serves as a means to lower the melting point of corundum, at the same time entering any fissures. The walls of the fissures are slightly melted, and upon cooling they are artificially healed, making the whole stone more stable. The flux is then trapped inside the fissures and shows as little droplets. Flux-healing is carried out in crucibles in muffled ovens at high temperatures. A centre for such heat treatments became Chantaburi in Thailand. Flux can only be detected when present on the surface of a gem, or if the gem had fissures and the flux entered the fissures. Usually a flux-healed gemstone has reached high enough temperatures to show other signs of heating, such as broken rutile needles or melted crystals, and often a chalky short-wave UV-fluorescence. In the mid-2000's, low-quality ruby was found in large amounts in East Africa, principally Madagascar and Mozambique. Such stones were not suitable for high-temperature heat treatment, because their fractured nature would have caused them to break. This led to the development of a new treatment with low temperature and a lead glass with refractive index similar to corundum. The lead glass enters the fissures, and the result is a transparent ruby. Once the glass solidifies, it also keeps the ruby in one piece, making it possible to cut larger gems. Lead glass is not stable to acids, so that the ruby may fall apart when exposed to such chemicals. Lead-glass filled ruby is on the market for prices only a fraction of

4.1 Treatments

what natural rubies are valued at and have already been encountered set in vintage jewellery mountings. Generally, lead glass treatment is easily detectable, because the infilled fissures reveal a blue to pink flash effect when viewed in the microscope, bubbles are seen on the fissures, and the lustre on the surface of the gemstone, when viewed in reflected light, is slightly different than the stone itself. Lead glass is also softer than corundum, so that it may show a slightly indented surface. Meanwhile, blue cobalt-lead glass is also used to improve the colour of sapphire.

During the heating process, chemical elements can be diffused and cause colour alterations on the gemstone (crystal lattice diffusion). Depending on the type of element, the diffusion may take place only close to the surface or may penetrate the gemstone completely. Lattice diffused corundum first appeared in the gem markets in 1979. Titanium, which, together with iron, causes the blue colour in sapphires, was introduced into the crystal lattice, and a uniform blue layer, several micrometers thick, surrounded the otherwise pale blue or colourless corundum. Since such gemstones have to be re-polished after diffusion, colour concentrations at the facet edges can often be observed. Similarly, chromium diffusion has been used to create rubies. Both treatments can also be detected by the higher than usual titanium- or chromium-concentrations during chemical analysis with XRF. However, while the diffusion with titanium is a widely used process still to date, chromium diffusion is much rarer and rarely found in the trade. In 2001, a new kind of orange coloured sapphire appeared in large quantities in the market. The colour cause was related to a complicated mechanism initiated by the addition of beryllium during the heating process. Beryllium penetrates the corundum deeper than titanium, making it nearly undetectable in the gemmological microscope. Only in some cases, where the diffusion did not penetrate the entire stone, an orange rim around a pink center can be seen in immersion. Since beryllium is a light element, it cannot be detected with X-ray fluorescence spectrometry, and therefore highly advanced methods such as LA-ICP-MS, LIBS, or SIMS have to be applied (Hanni and Pettke 2002; Hanni et al. 2004; Krzemnicki et al. 2004; Sastry et al. 2009).

A newer development is also the use of heat and pressure together on gems other than amber. Originally, high pressure high temperature (HPHT) autoclaves were built to produce synthetic diamonds as far back as the 1950s, but soon they were also used to treat natural diamonds of low quality (Elwell 1979). In the 1960s, Russian scientists experimented with treating natural diamonds, and in the late 1970s, researchers at GE (General Electric) obtained two U.S. patents on processes to remove yellow and yellow-brown colour from type I diamonds. In 1999, GE announced the use of HPHT treatment to remove colour from type IIa diamonds (nearly free of nitrogen), which had a greyish or brownish colour due to lattice distortions. The company announced that diamonds treated with this method were indistinguishable from natural ones. This led researchers to quickly find a method to detect this treatment (e.g., Chalain et al. 1999, 2000; Collins et al. 2000; Fisher and Spits 2000; Smith et al. 2000; Kiefert et al. 2005). The determination of a HPHT treatment is carried out by Photoluminescence Spectroscopy. Meanwhile, pressure-aided heat treatment is also applied on high quality tourmaline and on sapphire. However, the pressures for these treatments are not as high as for diamonds.

4.1.4 Irradiation of Gems

The onset of the twentieth century with the discovery of X-rays and radioactivity brought a whole new field of experiments on gems. High energy radiation (as X-rays) produces in most cases defects in the lattice of crystals. Defects selectively absorb the light, generating colours. First experiments showed that colourless diamond could turn green, yellow or blue, quartz turned brown, irradiation induced a yellow colour in corundum that produced yellow sapphires from colourless ones or green ones from blue ones, tourmaline could be turned red or yellow, topaz orange, and kunzite green. Most of these treatments were not stable and were reversed in light and/or heat. As more intensive sources of radiation became available, further reports occurred about irradiated gems (Pough 1957). By the 1970s, attractive blue beryl as well as blue topaz entered the market. While the blue "maxixe-type" beryl can be identified (Adamo et al. 2008; presence of blue "maxixe-type" beryl coloured by natural irradiation is still under discussion) by using some gemmological features (unusual dicroism with sometimes green fluorescence under ultraviolet, lack of iron absorptions in its visible spectra and presence of bands linked to irradiation instead). However, it is not possible to distinguish artificially irradiated topaz from natural blue topaz (i.e., naturally irradiated), except if there is residual radiation detected with a Geiger counter (present for the first and absent for the latter).

Irradiated and subsequently heated diamonds can be treated in various colours (Overton and Shigley 2008 and references therein). At the beginning when this treatment was used, often only a surface coloration was obtained, which could mostly be detected by microscopy. The irradiation caused a colourless diamond to change to green, however, since radioactive elements such as uranium often occur in the same geological environment as diamonds, green colour can also be caused by natural irradiation. Sometimes, a distinction between naturally or artificially irradiated green diamonds is not possible. Meanwhile, not only natural but also synthetic diamonds are routinely irradiated, and in combination with various heat treatment methods a range of colours including red, orange, yellow, brown and pink can be produced.

Freshwater pearls or saltwater pearls with a freshwater nuclei (bead) are darkened when exposed to irradiation. Radiation reacts with the manganese content which is relatively high for freshwater pearls and shells and darkens it. The degree of darkening is dependent on the intensity of the radiation used as well as the manganese content. In freshwater pearls the result can be so intense that it can imitate the colour of black Tahiti cultured pearls, whereas in saltwater pearls with bead there is only an effect when the nacreous layer is thin such as in Akoya cultured pearls, and usually such treated pearls have a grey or blue appearance after treatment. Natural-colour black pearls usually show different UV-Vis absorption bands and Raman spectra than irradiated coloured samples (Kiefert et al. 2001).

Irradiated topaz has been popular ever since and is still manufactured on an industrial level to date. Besides diamonds, topaz and pearls, today commonly irradiated gems comprise corundum (mainly yellow sapphires), morganite, kunzite, or quartz. Occasionally, irradiated emeralds were detected. However, the results in improving the colour is negligible.

4.2 Imitations

Just like with gem treatments, materials of lower value have been used to imitate more valuable gems since more than 4000 years back, and written reports and even recipes are available from that time such as in the Stockholm papyrus (*ca.* 300 CE), where detailed information about the falsification of pearls and other gems is given. And Pliny mentions that "To say the truth, there is no fraud or deception, which takes a higher profit than the forgery of gems". This sentence is still true to date. Some people distinguish imitations (natural materials) from simulants which are manufactured products. The oldest simulant was probably faience, a glazed-ceramic composition used in Egypt as early as 5000 BCE to imitate turquoise and other materials. Elwell (1979) describes a necklace featuring glazed blue steatite beads that was excavated at Badari in Egypt and dates to 4000 BCE, imitating Lapis Lazuli. Faience is considered a forerunner of transparent glass and was used widely in Egypt by 1000 BCE. It was used to imitate emerald, lapis lazuli, onyx, and turquoise among others. Noteworthy, some coloured glasses were considered as gems about 2000 years ago and before.

A reliquary cross and a monstrance from the Basel Cathedral treasure, both from around 1440, contain several paste/glass stones, most of them being blue, but also some yellow, one green and one colourless glass was encountered (Hanni et al. 1998; see also Chap. 5). Throughout the millennia, glass has been refined and improved so that it can imitate nearly every gem. The addition of lead gives a higher refractive index, imitating highly refractive gems such as ruby and sapphire. The addition of thallium leads to an increase in dispersion or "fire" (Nassau 1980). Recently, a nearly perfect glass imitation of tourmaline has been displayed at the Tucson Gem Show (Huber et al. 2017). Glass is softer than most gems and has an amorphous structure, making it singly refractive. The refractive index, however, varies widely depending on the composition. Glass usually also contains air bubbles and colour swirls which can easily be seen in the microscope, especially if it is an older glass. These properties, together with its softness and the singly refractive character, make it easy to distinguish glass from a natural gem.

Plastics came up with the invention of celluloid in 1855. Celluloid is a compound made of nitrocellulose and camphor, two natural substances. The first synthetic plastic was bakelite, which was patented in 1907. Since then, there has been a wide proliferation of different types of plastics, and many of them have been used as gem imitations. They include opaque materials to imitate turquoise, jade, ivory or coral, translucent imitations of tortoise-shell or pearls, and transparent plastics to imitate amber, ruby, emerald, amethyst, and even diamond. With the addition of a colouring agent, plastic can be dyed in any colour and to any effect. In the 1920's, Bakelite was discovered as a jewellery material, and highly regarded designers such as Coco Chanel included Bakelite bracelets and other pieces of jewellery in their costume jewellery collections. In 1928, Poly(methyl methacrylate) (PMMA), or acrylic glass, was invented, and first production with this material started in 1937. It is a transparent thermoplastic material and is strong, tough, and lightweight. In the

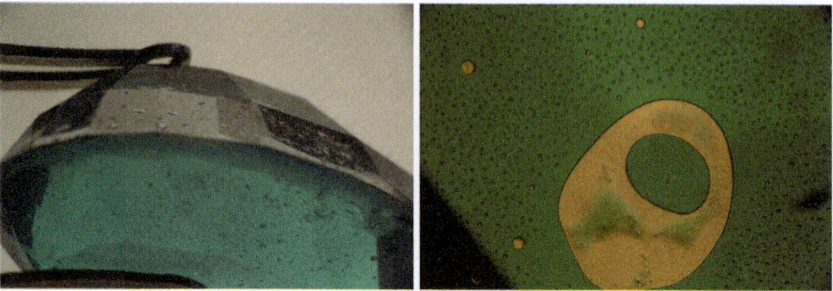

Fig. 4.8 Doublet consisting of natural quartz with fluid inclusions on the pavilion, a layer of green glue, and synthetic quartz on the top to imitate emerald. Field of view: 7 mm; left photo. In the microscope, the green glue is cleary visible. Size of the colourless bubble: 1.8 mm; right photo. (Photos: Lore Kiefert)

1950s and 1960s, Lucite, another trademark name for PMMA, was an extremely popular material for jewellery, with several companies specialized in creating high-quality pieces from this material. Lucite beads and ornaments are still sold by jewellery suppliers today. Today, plastics are also used to imitate fire opal and precious opal. The average specific gravity of plastics is around 1.05–1.55, and the general RI is about 1.5–1.6. Plastics are very soft (1.5–3 on Mohs' scale of hardness). They are sectile and will dissolve in some organic liquids. Most plastics will show some sign of degradation over time. A thermal reaction tester will produce an acrid smell. This test is widely used to distinguish plastic materials from natural amber, which has a much different aromatic smell.

Doublets usually consist of two different materials glued together (triplets if three materials are glued together), where the material on the top either is the colouring material and the bottom may be of a colourless material of lower value, of a coloured glass and a natural highly refractive mineral such as in garnet-topped doublets, or of two materials glued together with a coloured glue (Fig. 4.8). Triplets consist of three pieces, where the centre piece consists of the colouring layer. Doublets started to be very popular around 1850, but they were described already in The Mirror of Stones by Camillus Leonardus, MD in 1502: …"they fabricate the upper Superficies of the Granate, and the lower of Chrystal, which they cement with a certain Glew or Tincture; so that when they are set in Rings they appear like Rubies. …" and by Cellini (1568), who reports similar processes. In medieval religious objects, doublets are encountered quite frequently, such as in the reliquary cross of the Basel Cathedral from 1440, where several doublets with quartz top and red cement were identified, as well as one yellow glass doublet. In the monstrance of about the same age, the doublets with a quartz top were colourless and showed a decomposed cement, while the glass doublet was blue (Hänni et al. 1998). Garnet-topped doublets to imitate ruby were probably the most popular ones over the centuries, as ruby has always been one of the most expensive gemstones. However, doublets were also fabricated to imitate sapphire or emerald. Recently, a new generation of doublets to imitate emerald and other gemstones appeared on the market

4.2 Imitations

(Hanni and Henn 2015). Opal often comes in thin seams or slivers, and doublets or triplets are often encountered, where either the precious opal is backed by an opaque, mostly dark, common opal material (doublet), or the thin sliver is backed by the same material while the top consists of colourless quartz (triplet).

Doublets and triplets can usually be detected in the microscope as the layer of glue shows bubbles, or the top and bottom part are of different colour. Older garnet-topped doublets are often not mounted properly, so the interface can be observed on the crown as two materials with different reflection. When doublets and triplets are mounted in closed jewellery, it may be difficult to detect them if there are no bubbles or differences in inclusions, such as in garnet-topped doublets, which often contain needles in the garnet part but are inclusion free in the bottom part.

Since the Middle Ages, turquoise was sometimes imitated by heated fossilized mastodon ivory (a.k.a. odontolite). Fosilized mastodon ivory used for this purpose was 13–16 million years old (Miocene) and was principally found in southern France near to the Pyrenean mountain chain (Reiche et al. 2001; Krzemnicki et al. 2011). It is consisted of fluoroapatite ($Ca_5(PO_4)_3F$; dentine), i.e., differrent chemical formula than turquoise. During middle-ages Cistercian monks realized that by heating (over 600 °C), fossilized mastodon ivory was turning light blue and they believed they were creating turquoise. Several objects, principally religious, during and after this period were adorned with this material. In 1823 Fischer recognised that odontolite and turquoise are made of different minerals with different chemistry.

Jadeite "jade" may be imitated by ceramics, but a more common jadeite "jade" imitation is dyed quartzite. Quartzite is a metamorphic rock which is derived from sandstone by natural heat and pressure. The boundaries between the sand grains have the ability to take up fluids like dye. Dyed quartzite can easily be detected in the microscope due to its structure and the dye concentrations along grain boundaries.

Diamond possesses a high refractive index, high dispersion and the highest hardness found in any gemstone. Attempts to imitate diamond could very closely reproduce the dispersion and refraction, but never the hardness. Until the early twentieth century, when synthetic corundum entered the market, heat-treated colourless zircon was the most common diamond imitation. Synthetic spinel came on the market around 1935 (Read 2005), followed by synthetic rutile as diamond imitation in 1948, and strontium titanate soon thereafter. Garnet is singly refractive like diamond, and therefore diamond imitations like YAG (Yttrium-Aluminium-Garnet) or GGG (Gadolinium-Gallium-Garnet), which also have a much higher refractive index and dispersion than previous imitations, replaced the latter from approximately the 1970s on. At around the same time, first attempts to create Cubic Zirconia (CZ) were undertaken, and by the late 1970s, Cubic Zirconia (ZrO_2) replaced virtually all other materials as diamond imitation. Cubic Zirconia is still the most widely used diamond imitation today. In 1996, a new diamond simulant was introduced by C3 Incorporated of USA. The material is silicon carbide (SiC), called moissanite. The material has a hardness of 9.25, only slightly lower specific gravity (S.G.) and higher refractive index (R.I.) than diamond, and has a similar thermal conductivity. Especially this property makes it hard to distinguish from natural diamond, because thermal conductivity meters are widely used by jewellers to distinguish natural diamonds from imitations like Cubic Zirconia. However, it is strongly doubly refractive, and the doubling of facet edges is obvious when viewed in the microscope (Read 2005; Nassau et al. 1997; Kiefert et al. 2001).

Pearls are organic gems and have only been found naturally occurring principally in pearl producing bivalves. Since the 1920s pearls have been cultured in big scale, especially in Japan, and since that time culturing processes have been refined. Nowadays, natural pearls are considered extremely rare, and cultured pearls are most common. Despite the abundance of cultured pearls on the market, imitations can be found, especially in fashion jewellery. Shell beads, although of the same material as the pearl, never have the lustre of a pearl and are easily distinguished from natural pearls. Recently, beads from shells of *Tridacna* sp., a giant clam with thick inner part, were cut and used to imitate non-nacreous pearls. Glass beads with plastic covers can easily be detected due to their higher weight and the unnatural looking surface. Among the glass beads, Majorica imitation pearls are probably the most deceiving pearl imitations. They are manufactured by a meticulous process that involves multiple dipping of a glass nucleus into a compound extracted from fish scales. Each dipping is followed by a separate polishing of the bead. A final coating serves to harden the bead and protect it from discoloration by ultraviolet radiation (Hanano et al. 1990).

4.3 Synthetic Gemstones

Mankind was always fascinated by crystals and precious gems and attempts to synthesize them go back a very long time. The first record of crystal growth was the production of salt crystals by evaporating ocean water from about 2700 BCE on. In ancient India, sugar crystals were produced from a sugar solution, while Pliny mentions the crystallization of copper sulphate. By 1600, only two crystallization methods were known: evaporation and cooling of a solution. It took until the beginning of the nineteenth century, when crystallography was developed as a science, and another 100 years, when in 1902 the first synthetic gem-quality ruby appeared on the market in big quantities. Since that time, there was a rapid development in producing synthetic gemstones, starting with the Verneuil technique (flame fusion), to flux-grown synthetic gemstones, hydrothermal synthetics, high-pressure-high-temperature (HPHT) grown crystals, until the latest developments, chemical vapour deposition (CVD) crystals (Nassau 1980).

There are three major crystal-growth techniques: melt growth, solution growth and vapor phase growth. Several scientists such as Gaudin (1837), Böttger (1839) and Elsner (1839) could already produce tiny ruby crystals out of a potassium salt solution (Read 2005). In 1877, Frémy succeeded in using a lead flux (solution) to grow large numbers of very small ruby crystals. Some years later bigger ruby crystals were synthesized (a.k.a. Geneva or reconstructed rubies) using melt method and putting together small pieces of natural rubies; some scholars believe though that ruby powder was used. Large crystals were commercially obtained from ruby grown

4.3 Synthetic Gemstones

by the flame fusion technique by Verneuil, a former student of Frémy. In 1907, there was already a production of over 5 million of carats of ruby and sapphire. The flame fusion process (a type of melt growth) involves dropping powdered chemicals through a high-temperature flame, where it melts and falls onto a rotating pedestal to produce a synthetic crystal and forms a boule. Small inhomogeneities in the mixture lead to the formation of banding which is curved because of the shape of the boule. This banding is often the only microscopic evidence that it is a synthetic gemstone rather than a natural one. Adding a surplus of titanium oxide with subsequent annealing over a lengthy time produces a star effect (Hughes et al. 2017). Verneuil synthetic corundum was widely used in jewellery in the first half of the twentieth century, and for a short while was considered more valuable than its natural counterpart, because it was virtually inclusion-free. Today the Verneuil process remains the least expensive and most common way to make gems such as synthetic corundum and spinel. It can be frequently identified under microscope (Fig. 4.9).

The crystal pulling or Czochralski process (melt growth) also emerged in the early 1900s. In this process, nutrients are melted in a crucible where the synthetic crystals grow from a seed that is dipped into the melt and slowly pulled away from the melt as it grows. Gems made this way include corundum, synthetic alexandrite, chrysoberyl and garnet. Crystals grown by this method have a larger diameter and are cleaner than Verneuil synthetic gems. Colourless synthetic sapphires produced by this method are often used for watch glasses. Inclusion-free synthetic rubies, sapphires, and alexandrites can be produced by the floating zone melting process (also a melt growth technique), where a powder or partially fused powder moves through a zone of intense heat that melts the powder. After melting, impurities are carried out of the powder, and then recrystallized.

Flux-grown (solution growth technique) synthetic gems are grown from a melt, in which they grow on a seed crystal of the desired gemstone. Although there were earlier attempts, the techniques in use today are similar to the one originally developed in 1935 by IG Farben for emerald (Schmetzer and Kiefert 1998). In 1938, the American chemist Carroll F. Chatham started producing emerald on a commercial scale. Several other producers of emerald followed over the years, such as Pierre

Fig. 4.9 Curved striae in a Verneuil-synthetic colour-change sapphire. Field of view: 2.3 mm. (Photo: Lore Kiefert)

Gilson of France in 1963, Kyocera and Seiko in Japan, as well as several Russian companies. In 1959, the first flux-grown synthetic ruby from Chatham appeared on the market, followed by Kashan synthetic ruby in 1969, Knischka, Lechleitner and Ramaura in the 1980s, and Douros in 1993. Today, gemstones grown by this method encompass emerald, ruby, sapphire, alexandrite, and spinel. Flux-grown synthetic gemstones are much harder to detect than those grown by the Verneuil or the Czochralski process, because they form crystal faces and growth structures also encountered in natural gemstones. However, in some instances they develop additional crystal faces not encountered in natural stones. Other properties are flux residues, often noticed by their orange colour, and veil-like fingerprints which also can contain these residues. Hexagonal platinum platelets are found in several of the flux-grown gemstones. Chemical analysis often reveals the flux components or platinum from the crucible, and the FTIR spectra of flux-grown synthetic emeralds, for example, do not show any structural water bands. Hughes et al. (2017) give a detailed description of the various flux grown rubies together with the composition of the fluxes as well as detection methods.

Another method to produce synthetic gemstones is hydrothermal growth (also a solution growth technique). It grows crystals from an aqueous solution of the source material and requires heat and pressure in an autoclave. This method effectively duplicates the conditions in the earth that results in the formation of natural gems, especially quartz and emerald. The most widely used application for hydrothermal growth is the production of quartz. Colourless quartz crystals of up to 10 m in length are mentioned (O'Donoghue 2005), but synthetic amethyst, smoky quartz, citrine and other colours usually occur in slabs of about 20–25 cm in length (Balitsky et al. 1999). Synthetic colourless quartz is mainly used for the watch industry because of its piezoelectric properties, while during the 1980's, it was reported that synthetic amethyst accounted for up to 25% of amethyst carried by dealers in the Far East (O'Donoghue 2005). Meanwhile, the production of hydrothermal synthetic quartz and its varieties has increased even more, and it is virtually impossible nowadays to separate high quality natural from synthetic quartz without advanced and therefore expensive testing methods (Karampelas et al. 2011).

The other gem most successfully produced by the hydrothermal method is emerald (Fig. 4.10). While flux-grown emeralds have been on the market much longer, they usually take longer to grow and yield smaller crystals. J. Lechleitner of Austria applied this process in 1960 to create a synthetic overgrowth over natural beryl, but later also produced full synthetic emeralds. From 1965, the Linde Division of Union Carbide (later Regency) produced large volumes of hydrothermal synthetic emeralds, followed by Russia in 1966, Biron in Australia in 1987, and later by Japan's AGEE (Hanni and Kiefert 1994) as well as Chinese producers (Schmetzer et al. 1997). Hydrothermal synthetic emeralds have typical "chevron-type" growth structures, which make them microscopically distinguishable from their natural counterparts in most cases (Fig. 4.11). Even Raman spectroscopy can be helpful in the identification of hydrothermal emeralds, showing the typical sharp band of the "alkali-free" water vibration (Bersani et al. 2014). Other synthetic beryl such as aquamarine or pink beryl is also produced with this method.

4.3 Synthetic Gemstones

Fig. 4.10 Hyrdrothermal synthetic emerald of 3 cm height. (Photo: Stefanos Karampelas/LFG)

Fig. 4.11 So-called "chevron"-like growth features in a hydrothermal synthetic emerald. Field of view: 4 mm. (Photo: Natalya Hermenau, GGL)

Hydrothermal synthetic corundum is not as common as flux-grown synthetic corundum; however, they are in production in Russia since 1993 (Peretti and Smith 1993; Peretti et al. 1997). Hydrothermal synthetic corundum has similar chevron-like growth zoning as hydrothermal synthetic emerald and should therefore be easily distinguishable from their natural counterparts (Schmetzer and Peretti 1999).

The first attempts to synthesize diamonds was undertaken in 1953 by a Swedish company. But it took until 1970, when synthetic gem quality diamonds up to carat-sizes were produced by a high-pressure/high-temperature (HPHT) process (flux grown method). They are grown from a melt flux which dissolves carbon at higher temperatures, which then, together with pressure, converts the hexagonal atomic structure of graphite into the more tightly bonded cubic structure of diamonds (Read

2005). In 1986, Sumitomo Electric Industries of Japan announced the commercial production of HPHT yellow synthetic diamonds of gem quality, and in 2004, Chatham Created Gems in the USA began marketing a series of new HPHT synthetic diamonds in a range of colours, including near-colourless diamonds. The first diamonds were easily distinguishable by their distinct growth pattern around a seed as well as with spectroscopic methods (Shigley et al. 1994 and references therein). Meanwhile, pure colourless diamonds with sizes over 5 cts can be produced by HPHT methods, and their detection is mostly restricted to spectroscopic features (Poon et al. 2015).

Chemical vapour deposition (CVD) synthetic diamonds (synthesized by vapor phase growth) had their breakthrough in the early 1980s and experienced a "pre-commercial" phase until 2008. Since then, CVD synthetic diamonds have improved and have gone in commercial production, but it is only since 2013 that they are more frequently encountered in the gem trade. The process involves diamond growth at moderate temperatures (700–1300 °C) in a mixture of a hydrocarbon gas and hydrogen. Within a vacuum chamber, activation of the gas by an energy source (typically a microwave plasma) breaks apart the gas molecules to release carbon atoms. These are drawn to a substrate consisting of synthetic diamond forming diamond tablets. CVD synthetic diamonds have only few inclusions which consist mainly of clouds, pinpoints and black graphite crystals, and display an orange to red UV fluorescence in pink samples. Under crossed polarizers they display a strong strain pattern, but otherwise the detection has to be undertaken with spectroscopic methods (Eaton-Magana and Shigley 2016).

Other synthetic gems comprise coral (ceramic), turquoise (ceramic), lapis lazuli (ceramic) and opal (aqueous solution), whereas most of these are easily recognized. Synthetics to imitate diamond have been mentioned above, such as Cubic Zirconia (by melt technique; skull melt), YAG (by melt technique; Czochraski), GGG (by melt technique; Czochraski) and moissanite (by vapor technique).

4.4 Final Remarks

Although this chapter has tried to cover all relevant gem treatments, imitations and synthetics, the list is by far not complete, and one always has to be aware that nearly all materials can be altered or imitated. However, some of the treatments and synthetics are very young and it is questionable if they play a role for the archaeologist or art historian unless it is a reproduction. Some gems, such as Paraiba tourmaline or tanzanite, were only found after the middle of the last century, as well as some ruby and sapphire sources. Synthetic diamonds, for example, should not be a subject of concern when dealing with historic pieces.

References

Abduriyim A, Kimura H, Yokoyama Y, Nakazono H, Wakatsuki M, Shimizu T, Tansho M, Ohki S (2009) Characterization of green amber with infrared and nuclear magnetic resonance spectroscopy. Gems Gemol 45:158–177

Adamo I, Pavese A, Prosperi L, Diella V, Ajò D, Gatta GC, Smith CP (2008) Aquamarine, maxixe-type beryl, and hydrothermal synthetic blue beryl: analysis and identification. Gems Gemol 44:214–226

Balitsky VS, Lu T, Rossman GR, Makhinova IB, Mar'in AA, Shigley JE, Elen S, Dorogovin BA (1999) Russian synthetic ametrine. Gems Gemol 35:122–134

Bauer M (1896) Ueber das Vorkommen der Rubine in Birma. Neues Jahrbuch fur Mineralogie, Geologie und Palaeontologie 2:197–238

Bersani D, Azzi G, Lambruschi E, Barone G, Mazzoleni P, Raneri S, Longobardo U, Lottici PP (2014) Characterization of emeralds by micro-Raman spectroscopy. J Raman Spectrosc 45:1293–1300

Chalain JP, Fritsch E, Hänni H (1999) Detection of GE POL diamonds: a first stage. Revue de Gemmologie AFG 138(/139):30–33

Chalain JP, Fritsch E, Hänni H (2000) Identification of GE POL diamonds: a second step. J Gemmol 27:73–78

Collins AT, Kanda H, Kitawaki H (2000) Colour changes produced in natural brown diamonds by high-pressure, high-temperature treatment. Diam Relat Mater 9:113–122

Eaton-Magana S, Shigley JE (2016) Observations on CVD-grown synthetic diamonds: a review. Gems Gemol 52:222–245

Elen S (2001) Spectral reflectance and fluorescence characteristics of natural-color and heat-treated "golden" south sea cultured pearls. Gems Gemol 37:114–123

Elwell D (1979) Man-made gemstones. Ellis Horwood, Chichester, 191 pp

Fisher D, Spits RA (2000) Spectroscopic evidence of GE POL HPHT-treated natural type IIa diamonds. Gems Gemol 36:42–49

Gubelin EJ (1950) New process of artificially beautifying gemstones. Gems Gemol 6:243–254

Gubelin E, Graziani G, Kazmi AH (1986) Pink Topaz from Pakistan. Gems Gemol 22:140–151

Hanano J, Wildman M, Yurkiewicz PG (1990) Majorica imitation pearls. Gems Gemol 26:178–188

Hanni HA (1992) Identification of fissure-treated gemstones. J Gemmol 23:201–205

Hanni HA, Henn U (2015) Modern doublets, manufactured in Germany and India. J Gemmol 34:479–482

Hanni HA, Kiefert L (1994) AGEE hydrothermal synthetic emeralds. Jewel Siam Oct/Nov: 80–85

Hanni HA, Pettke T (2002) Eine neue Diffusionsbehandlung liefert orangefarbene und gelbe Saphire. Zeitschrift der deutschen Gemmologischen Gesellschaft 51:137–152

Hanni H, Schubiger B, Kiefert L, Häberli S (1998) Raman investigation on two historical objects from Basel cathedral: the reliquary cross and Dorothy monstrance. Gems Gemol 34:102–125

Hanni HA, Krzemnicki MS, Kiefert L, Chalain JP (2004) Ein neues Instrument für die analytische Gemmologie: LIBS. Zeitschrift der deutschen Gemmologischen Gesellschaft 53:79–86

Huber B, Kiefert L, Link K, Laurs B (2017) Borosilicate glass resembling gem crystals. J Gemmol 35:494–496

Huda SNA (1998) Arab roots of gemology: Ahmad ibn Yusuf al Tifhashi's best thoughts on the best of stones. Scarecrow Press, Lanham, 274 pp

Hughes RW, Emmett JL (2005) Heat seeker. Guide 24(1):4–7

Hughes RW, Manorotkul W, Hughes EB (2017) Ruby & sapphire: a gemologist's guide. Lotus Publishing, Bangkok, 816 pp

Karampelas S, Fritsch E, Zorba T, Paraskevopoulos KM (2011) Infrared spectroscopy of natural vs. synthetic amethyst: an update. Gems Gemol 47:196–201

Kiefert L (2015) Natural green Amber from Ethiopia. In: Proceedings of the 34th international gemmological conference 2015, Vilnius, Lithuania, pp 22–25

Kiefert L, Schmetzer K, Hänni HA (2001) Synthetic moissanite from Russia. J Gemmol 27:471–481

Kiefert L, Chalain JP, Häberli S (2005) Case study: diamonds, gemstones and pearls: from the past to the present. In: Edwards HGM, Chalmers JM (eds) Raman spectroscopy in archaeology and art history, vol XXI. Royal Society of Chemistry, Cambridge, pp 379–402

Konig R, Hopp J (eds) (1994) C. Plinius Secundus d. Ä., Naturkunde. Lateinisch-deutsch, Buch XXXVII: Steine: Edelsteine, Gemmen, Bernstein. Artemis & Winkler, Zürich, 263 pp

Krzemnicki MS, Hänni HA, Walters RA (2004) A new method for detecting be diffusion-treated sapphires: laser-induced breakdown spectroscopy (LIBS). Gems Gemol 40:314–322

Krzemnicki KS, Herzog F, Zhou W (2011) A turquoise jewelry set containing fossilized dentine (odontolite) and glass. Gems Gemol 47:296–301

Misiorowski E (2000) Pretty in pink. Professional Jeweller, January 2000

Nassau K (1980) Gems made by man. Chilton Book Company, Radnor, 364 pp

Nassau K (1984) Gemstone enhancement. Butterworths, London, 221 pp

Nassau K (1985) Altering the color of topaz. Gems Gemol 21:26–34

Nassau K, McClure SF, Elen S, Shigley JE (1997) Synthetic moissanite: a new diamond substitute. Gems Gemol 33:260–275

O'Donoghue M (2005) Artificial gemstones. NAG Press, London, 294 pp

Overton TW, Shigley JE (2008) A history of diamond treatments. Gems Gemol 44:32–55

Peretti A, Smith CP (1993) A new type of synthetic ruby on the market: offered as hydrothermal rubies from Novosibirsk. Aust Gemmol 18:149–157

Peretti A, Mullis J, Mouawad F, Guggenheim R (1997) Inclusions in synthetic rubies and synthetic sapphires produced by hydrothermal methods (TAIRUS, Novosibirsk, Russia). J Gemmol 25:540–561

Poon PY, Wong SY, Lo C (2015) Large HPHT-grown synthetic diamonds examined in GIA's Hong Kong laboratory. Gems Gemol 51:65–66

Pough FH (1957) The coloration of Gemstones by Electron Bombardment. Sonderheft zur Zeitschrift der Deutschen Gemmologischen Gesellschaft für Edelsteinkunde, pp 71–78

Read PG (2005) Gemmology, 3rd edn. Elsevier, Oxford, 324 pp

Reiche I, Vignaud C, Champagnon B, Panczer G, Brouder C, Morin G, Sole VA, Charlet L, Menu M (2001) From mastodon ivory to gemstone: the origin of turquoise color in odontolite. Am Mineral 86:1519–1524

Sastry MD, Mane SN, Gaonkar MP, Bagla H, Panjikar J, Ramachandran KT (2009) Evidence of colour-modification induced charge and structural disorder in natural corundum: spectroscopic studies of beryllium treated sapphires and rubies. IOP Conf Ser Mater Sci Eng 2:1–4

Schmetzer K, Kiefert L (1998) The colour of Igmerald (I.G. Farbenindustrie flux-grown synthetic emerald). J Gemmol 26:145–155

Schmetzer K, Peretti A (1999) Some diagnostic features of Russian hydrothermal synthetic rubies and sapphires. Gems Gemol 35:17–28

Schmetzer K, Kiefert L, Bernhardt HJ, Zhang B (1997) Characterization of Chinese hydrothermal synthetic emerald. Gems Gemol 33:276–291

Shigley JE, Fritsch E, Koivula JI, Sobolev NV, Malinovsky IY, Pal'yanov YN (1994) The gemological properties of Russian gem-quality synthetic yellow diamonds. Gems Gemol 29:228–248

Smith CP, Bosshart G, Ponahlo J, Hammer VMF, Klapper H, Schmetzer K (2000) GE POL diamonds: before and after. Gems Gemol 36:192–215

Themelis T (1992) The heat treatment of ruby and sapphire. Gemlab Inc, Bangkok, 236 pp

Themelis T (2004) Flux-enhanced rubies & sapphires. Gemlab Inc, Bangkok, 48 pp

Wang W, Scarratt K, Emmett JL, Breeding, Douthit CR (2006a) The effects of heat treatment on zircon inclusions in Madagascar sapphires. Gems Gemol 42:134–150

Wang W, Scarratt K, Hyatt A, Shen A, Hall M (2006b) Identification of "chocolate pearls" treated by Ballerina Pearl Co. Gems Gemol 42:222–235

Webster R, Anderson BW (1983) Gems: their sources, description and identification, 4th edn. Butterworths, London, 1006 pp

Chapter 5
Archaeometrical Questions (Case Studies)

The majority of gems of archaeometrical interest are focused are mounted, so the possible methods to use are restricted. Additionally, for security reasons most of these items cannot be tested outside of their host place, consequently only the instruments available there or mobile instruments can be used. It is sometimes interesting, as well as useful, to have some supporting documents on the studied items, such as inventory lists of museums with information on their origins, or invoices of previous sales. The current chapter presents examples gem analysis with answers to archaeometrical questions such as gem and imitation identification, provenance of gems as well as age of pearls and whether two historic diamonds were cut from the same rough crystal or not.

5.1 Identification of Gems

A group of exceptional objects belonging to the treasury of Einsiedeln Abbey, an important Benedictine monastery in Einsiedeln, Switzerland, was studied to identify the materials used in their construction. Between others, one ciborium made in 1592 with 17 coloured gems (Fig. 5.1) and one chalice made in 1609 with 28 coloured gems and 24 pearls, were studied in more detail in order to identify them (Fig. 5.2). The items were loaned to the Swiss National Museum for identification of the materials used in their construction. The results were compared with the observations made by Father Eustache Tonassini from 1794 to 1798, during the documentation of the treasures of Einsiedeln Abbey (Karampelas et al. 2010, 2012). For security reasons, the items could not be removed from the Swiss National Museum laboratory; thus, all testing took place there using the available instruments. Additionally, all gems were held in closed-back settings which was a further restriction in studying these objects. Microscopic examinations and UV fluorescence using a long-wave 365 nm and short-wave 254 nm lamp were performed on all stones. The geometry of the objects permitted EDXRF analysis on 6 coloured

---

The original version of this chapter was revised: Figures are updated without changing their captions. The correction to this chapter is available at https://doi.org/10.1007/978-3-030-35449-7_6

Fig. 5.1 This gold ciborium (33 cm high) adorned with coloured gemstones, was made at around 1592 and is part of the treasures of Einsiedeln Abbey in Switzerland. (Photo: Swiss National Museum)

gems on the ciborium as well as another 6 gems and 15 pearls from the chalice. It was possible to study all the gems and pearls with a Raman spectrometer, an L-shaped lens (magnification 30×), and a camera for adequate positioning of the beam, using three different lasers (532, 633 and 785 nm) in order to avoid or decrease the undesired luminescence (Fig. 5.3). 17 coloured stones of the ciborium (10 pinkish red, 4 orange, and 3 light blue) and the 28 coloured stones of the chalice (17 pinkish red to red, 6 purple, 2 orange, 1 light blue, 1 light yellow and 1 yellow green) were polished (some with simple cuts); they measure from 3.5 to 15.2 mm in length. No indications of doublets, imitations, glass, or synthetics and any indication of treatment were observed with magnification.

In the ciborium, using Raman spectroscopy, it was found that all 10 pinkish-red gems, mentioned to be ruby in the manuscript, were garnets. The main Raman band at 920 cm^{-1} indicates Al-garnets of the pyralspite series (pyrope-almandine-spessartine). Semi-quantitative calculations of their approximate composition on the basis of the Raman spectra, have shown that they consist of 58–75% almandine with 25–38% pyrope and some of them contain 2–4% andradite and/or spessartine (Bersani et al. 2009). With magnification, these garnets showed mainly needle-like inclusions, probably rutile, and inclusions with tension haloes, probably zircons. All 4 orange stones were Ca-garnets with the main Raman band at around 880 cm^{-1}.

5.1 Identification of Gems

Fig. 5.2 This gold chalice (21.2 cm high) adorned with coloured gemstones, was made at around 1609 and is part of the treasures of Einsiedeln Abbey in Switzerland. (Photo: Swiss National Museum)

Fig. 5.3 A Raman spectrometer with a L-shaped lens was used to take spectra of a gem difficult to access with conventional equipment. (Photo: Swiss National Museum)

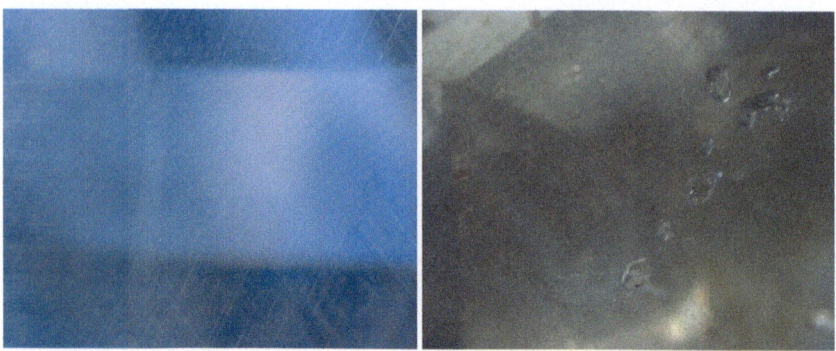

Fig. 5.4 Needle-like rutile inclusions (left) and multiphase inclusions (right) of two sapphires from the ciborium of Fig. 5.1. Field of view: about 0.9 mm (left) and 1.5 mm wide (right). (Photo: Swiss National Museum)

Semi-quantitative calculations have shown that they are 96–98% grossular and contain 2–4% pyrope. Under the microscope they contained negative crystals (i.e., cavities within a crystal) and healed fissures (i.e., fissures filled with natural material -most commonly fluid- during or after the growth of the gem). These were identified by Father Tonassini as "hyacinth"; a term which described different orange to yellow gems (see Chap. 2), among them yellow to orange to brown coloured zircon and grossular garnets (a.k.a. hessonite). All the garnets (almandine and grossular) were inert to UV radiation. The three blue stones presented the main corundum Raman band at about 415 cm^{-1} and a less intense band at 645 cm^{-1}. When changing the laser chromium photoluminescence bands were observed (e.g., at 1370 and 1400 cm^{-1} with 633 nm excitation and 4365 and 4395 cm^{-1} with 532 nm excitation which is at around 693 nm and 694 nm). These three stones were correctly described as blue sapphires. Under the microscope thin and long needles, probably rutile, negative crystals, multiphase inclusions and black particles were observed (Fig. 5.4). They fluoresced faint orange-yellow to long-wave UV and faint orange to short-wave UV. Taking into account when these stones were set and the oriental origin mentioned in the manuscript, Sri Lanka was the most probable source of these sapphires. In addition, their inclusions were consistent with sapphires from this origin. At the time, garnets were also known to come from the same locality (as well as from India), however other sources of pink to red garnets were also known back then.

The chalice was decorated with 24 baroque and button-shape pearls. The pearls were white to very light grey in colour, mostly button-shaped and baroque and measuring from 4.4 to 7.7 cm in diameter. Black spots, probably due to the decomposition of the pearl organic matters during time ('ageing'), were observed virtually on all the pearls. All pearls revealed typical growth lines ("fingerprints" like structure due to overlapping aragonite platelets) under high magnification, which are caused by the stacking of aragonite platelets. The growth lines are also known as 'steps', 'fingerprints' or 'contour lines' in geological and geographical maps. These gave clear indications that they are indeed pearls and not imitations. Raman spectra on five of the pearls presented aragonite bands. Under long-wave UV radiation, the pearls fluoresced yellowish green, and under short-wave, they fluoresced green.

EDXRF analyses mainly detected Ca due to the fact the pearls are made from calcium carbonate, with small amounts of SrO and very low amounts (some below detection limit) of MnO. By plotting the results to a binary diagram MnO/SrO and by comparing them with the data acquired using the same measuring conditions on reference freshwater and saltwater samples, the pearls appeared to be of saltwater origin. By considering the manufactured time of the object, as well as their size and colour, the Arabian Gulf, Red Sea and Gulf of Mannar (between India and Sri Lanka) were the most probable origins of these pearls.

Raman spectra on 15 of the 17 pinkish red to red stones, mentioned to be ruby in the manuscript, presented high luminescence under the three excitation wavelengths. By using the 785-nm laser, the main corundum bands were visible. Under short-wave and long-wave UV lamp excitations, all rubies presented medium intense to intense red colour fluorescence. This was probably due to their relatively high chromium and low iron content, characteristic of marble hosted rubies. Moreover, all 15 rubies contained thick and short needle-shaped inclusions -probably rutile-, negative crystals as well as colourless and red colour zoning. The characteristics of these rubies was consistent with those observed in some rubies from Myanmar. Thus, taking this into consideration as well as the period during which these stones were set (and assuming that they were not changed in the meantime) and oriental origin mentioned in the manuscript, it is most probable that they were mined from the Mogok mines in Myanmar (Burma). Raman spectra on the 2 biggest of the 17 pinkish red to red stones (one mentioned to be ruby in the manuscript and the other mentioned to be either ruby or garnet), showed the main Raman band around 920 cm^{-1} and semi-quantitative calculations of their approximate composition have shown that they are both almandine garnets (62% and 74% respectively).

The light blue stone presented also very similar features to those of the light blue sapphire in the ciborium. Three stones, two orange and one light yellow, were described in the manuscript as topaz. Noteworthy, the name topaz was affixed to a mineral with chemical formula $Al_2SiO_4(F,OH)_2$ and before was mainly referred to peridot (gem quality yellowish-green to green olivine $((Mg,Fe)_2SiO_4)$). Raman spectra on the two orange stones showed the main Raman band at around 880 cm^{-1} and a series of bands below, indicative for Ca-garnets and semi-quantitative calculations have shown that these gemstones contain 90–96% grossular. For the above, classical gemmological observations (fluorescence and microscopy) were similar to those observed in the ciborium; thus similar provenances are possible.

The light-yellow stone showed an intense band at about 464 cm^{-1} with less intense bands at lower Raman shifts; a spectrum characteristic of quartz. Thus, this stone was classified as citrine (yellow variety of quartz). The stone was also inert under UV excitation and presented some fissures and fluid inclusions under the microscope. The Raman spectra on the six purple stones were similar to those obtained on the citrine described previously. They were inert under UV excitation and also presented similar inclusions to the citrine. All six stones were amethysts (purple variety of quartz), correctly mentioned in the manuscript. Gem quality citrine and amethyst can be found in many deposits around the world. The yellow-green stone of this object was described in the manuscript as chrysolith, another term for peridot (gem-quality olivine). Raman spectra on this stone presented a double peak at about 824 and 855 cm^{-1} characteristic of olivine, as well as other olivine related bands with an approximate composition of 92% forsterite (i.e., Mg-rich olivine). Under UV light, the stone was inert, and under the microscope, it was relatively clean with some tiny fluid inclusions. Peridot from Zabargad island (this island is also cited in old texts under other names such as topazios) in the Red Sea (at the off-coast Egypt) was known for more than 2000 years. However, other sources including Saudi Arabia, Pakistan, Sri Lanka, Myanmar and China are known today without having all the needed information of when exactly these mines started to be exploited.

5.2 Identification of Imitations

During the investigation of two ecclesiastic items (The Reliquary Cross and Dorothy Monstrance), of the late Gothic period (1350–1520), from the treasury of Basel Cathedral (Basel Munster, Basel, Switzerland), some doublets as well as variously coloured glasses were found along with natural gems (Hanni et al. 1998). All gems had a closed back setting and they were studied under the microscope, with a hand-held spectroscope as well as a Raman spectrometer. With few exceptions, all gems in both objects were cut in symmetrical shapes (e.g., oval, octagonal or rectangular). The style of cutting usually uses a slightly domed table and one step of parallel facets on the crown.

The Reliquary Cross contains 28 stones, only 16 were natural with 15 different coloured quartz (8 amethyst, 4 colourless quartz, 2 smoky quartz and 1 citrine) with multiphase inclusions, and 1 turquoise (Fig. 5.5). The other 12 stones are doublets and glass imitations. However, because of their mounting it was not possible to identify the pavilion material of the doublets. The four red coloured doublets found to have quartz tops with a cement layer that showed a red pigment mixed into the adhesive. Most cement layers appeared to have dried out in most of the doublets probably due to their age. The other 4 yellow and 4 blue stones were made out of glass, with round and elongated air bubbles. With the hand spectroscope, a weak cobalt spectrum was observed in the blue stones.

The Dorothy Monstrance was adorned with 19 stones but only 8 were natural (Fig. 5.6). These consisted of peridot, blue sapphire, red garnet, pink spinel, amethyst, and three cryptocrystalline forms of silica (quartz) such as a carnelian (orange coloured chalcedony) finely engraved with a goat, a black onyx (black coloured chalcedony), and an onyx (black and white banded agate) engraved with the profile of a woman's head. The rest were both quartz doublets (with heavily decomposed cement layers) and glass. All identifications were also confirmed by Raman spectroscopy.

Fig. 5.5 The Reliquary Cross of the Treasure of the Basel Cathedral, believed to have been manufactured around 1440. Height of the cross: 37 cm. (Photo: P. Portner, courtesy of Historisches Museum Basel)

Geographic origin of quartz varieties and cryptocrystalline silica was difficult as in central Europe, several occurrences of such stones have been known for centuries. Similarly, for red garnets of this period many sources were known. However, the spinel and sapphire showed inclusions and growth zoning that suggested a Sri Lankan origin. The turquoise and peridot probably originated from one of the Near East deposits (Nishapur/Khorassan region, Persia; and Zabargad, Egypt, respectively), since the pieces pre-date the discovery of America.

Studies with mobile Raman spectrometers for the identification of gems in jewellery pieces have revealed that glass is frequently present; as for example in items from the collection of the Interdisciplinary Regional Museum Maria Accascina, Messina, Sicily, Italy (Barone et al. 2015). Some of these stones were misidentified previously: a couple of examples are the yellow glasses previously identified as topaz in a half-moon shaped choker #A 96a (the second and fifth big stones counted from left to right in Fig. 5.7 left, the remaining four are quartz), and a red glass previously identified as ruby in the hair clip #108 (the biggest red stone in Fig. 5.7 right, the other red stones are actually rubies). In another item back painting, possibly to alter gem's colour, was identified by Raman spectroscopy (Fig. 5.8).

Some gem treatments have been documented since antiquity; for instance, heating of quartz, colouring of cryptocrystalline silica varieties, oiling of emeralds (see chapter on treatments), however, virtually no studies on treated gems used in jewellery of archaeological interest were published. This is likely due to the fact that the

Fig. 5.6 The Dorothy Monstrance of the Treasury of Basel Cathedral, believed to be of the same time as the Reliquary Cross of Fig. 5.5, is 55 cm high. (Photo: P. Portner, courtesy of Historisches Museum, Basel)

Fig. 5.7 Half-moon shaped choker (seventeenth century) with yellow glasses imitating topaz gems (length of the choker 11 cm) on the left and hair clip (seventeenth century) with a red glass imitation between natural rubies (length of the clip 7 cm) on the right. (These items are courtesy of Regione siciliana, Assessorato dei Beni Culturali e Identità siciliana, Dipartimento dei Beni Culturali e della Identità siciliana – Museo Regionale Interdisciplinare di Messina – Messina, Sicily, Italy)

5.2 Identification of Imitations

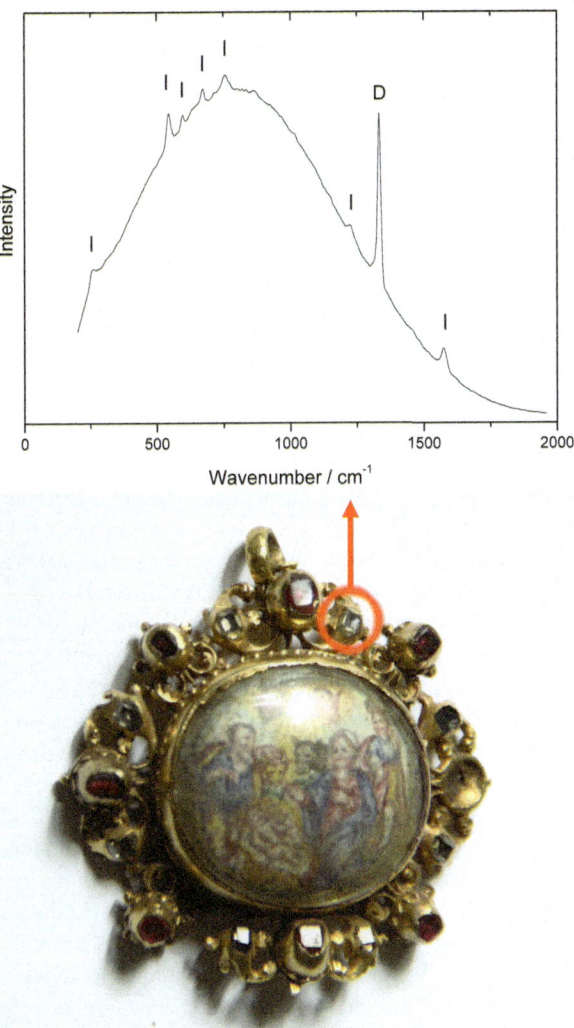

Fig. 5.8 Raman spectrum performed on a diamond present in a seventeenth century jewel (height 4.5 cm) revealing the presence of an indigo backpainting (*D* diamond, *I* indigo). (This item is courtesy of Regione siciliana, Assessorato dei Beni Culturali e Identità siciliana, Dipartimento dei Beni Culturali e della Identità siciliana – Museo Regionale Interdisciplinare di Messina – Messina, Sicily, Italy)

researchers who were studying these gems were not familiar with gem treatment identification, so treatments of gems might have been overlooked.

5.3 Provenance

Geologic and geographic origin determination could be helpful to trace and/or reconsider some trade routes, as well as provide a better understanding of the history of the studied sample with reference to the content of old books. In rare gems such as blue sapphires, rubies and emeralds, information specific to their provenance can be easier collected compared to quartz and garnets. This is because the first are found in much more specific geological environments than the second. Origin determination is actually based on small differences in trace element composition, in spectroscopic features as well as in inclusions. Usually, it requires sophisticated (and expensive) laboratory techniques along with an experienced gemmologist.

Important work on origin determination of ancient emeralds was done in the course of the last 20 years using oxygen isotopes, hydrogen isotopes, chemical and spectroscopic analysis as well as inclusions (Giuliani et al. 1998, 2000, 2001; Rondeau 2003; Schwarz and Pardieu 2009; Smith and Darenius 2009; Groat et al. 2014; Saeseaw et al. 2014). Noteworthy, that frequently historical records helped to draw the conclusion on the gems provenance. An example was presented on four emeralds from the Museum National d'Histoire Naturelle in Paris, France. One sample was the St. Louis Emerald, a translucent cabochon (rounded rectangular), measuring 30 × 23 × 7 mm, with an estimated weight of 51.6 ct. No historical record was found on this sample. Some studies from the museum curators comparing it with paintings and sketches, taking into consideration the dimensions of the emerald and face characteristics (nose and eyes) of Saint-Louis (Louis IX; 1214–1270) with his crown have shown that the green stone in the frontal part of certain paintings has similar shape and dimensions with the studied emerald. A translucent emerald of 1.21 cts, 6 × 4.5 mm hexagonal shape (very similar to the rough beryl shape) was mounted on a yellow metal wire, believed to be part of an earring. It was drilled along the elongated axis. This object was discovered during an archaeological excavation in Eastern France, near Miribel (Ain department near Lyon) in a cellar dated back to the Gallo-Roman time period (i.e., mid-first century BCE to end-fifth century CE); without more information. The other two emeralds were also translucent and weighed 12.57 and 10.9 ct with dimension of 14.2 × 13.5 mm and 15 × 12 mm respectively. They were both rounded rectangular cabochons and were belonging to the mineral collection of mineralogist, René-Just Haüy (1743–1822). In the notes found with the collection, which were not written by him, it was mentioned that these samples were from Egypt. Oxygen isotope analysis $^{18}O/^{16}O$ ($\delta^{18}O$) on the four samples were compared with those acquired on emerald samples from known origins which ranged from 6.2 parts per thousand (‰; SNOW: Standard Mean Ocean Water) to 24.7‰. Three micro-destructive analyses of the St Louis

5.3 Provenance

emerald and one on the biggest sample from the Haüy collection gave $\delta^{18}O$ of 7.5 ± 0.5‰ and 7.6‰. This data was consistent with the data acquired on the reference samples from Austria and Zimbabwe and the first from Brazil as well. Brazilian emeralds arrived in Europe after the sixteenth century (discovery of Brazil by the Portuguese) and emerald mines from Zimbabwe were only known since the mid-1960s. On the other hand, Austrian emeralds are known for several centuries. Taking into account that the first emerald adorned the St. Louis crown (thirteenth century); Zimbabwe and Brazil cannot be an option. This is also true for the biggest sample from the Haüy collection (early nineteenth century the latest), Zimbabwe was excluded, and Austria was suggested as the most probable source. The observed inclusions were also in agreement with the above conclusion (Giuliani et al. 2000). The other emerald from the Haüy collection presented significantly higher values; $\delta^{18}O$ of 10.5‰; indication that it is not coming from the same mine. This value was suggested to be consistent with the reference samples from Egypt (Giuliani et al. 2000). However, later reference samples from other mines (e.g., India) gave similar values and can, therefore, also be possible sources of this sample. Two analyses on the sample from the Gallo-Roman period presented $\delta^{18}O$ of 15.2 ± 0.3‰ and were consistent with the reference samples from Pakistan. Some samples from Colombia may also present similar values, however these were not known in Europe during that time. This means that emeralds, as well as other goods, were transported from Pakistan to Europe for probably more than 2000 years and it was suggested that Bactrian emeralds cited by Theophraste are possibly coming from Pakistan.

Provenance studies were also done by PIXE on three ruby inlays of a Parthian statuette (third century BCE) originating from Babylon, Mesopotamia (some of the oldest rubies ever identified; Calligaro et el. 1998). The results of the samples were compared with those acquired on the reference ruby samples from different origins. Based on the relatively low amount of iron and relatively high amount of vanadium, the authors concluded that the samples can possibly come from Sri Lanka or from Burma. Bibliographic references of Sri Lankan rubies from this period of time were found. Apart from the chemistry, spectroscopic data as well as inclusions can also give valuable information about the provenance of a coloured stone. Some examples on the provenance and inclusions found in blue sapphires and rubies were also described in the sections above.

Another important example is the 135.74 ct Grand Sapphire, a sapphire added to the French Crown Jewels of Louis XIV in 1669 (Farges et al. 2015). The possible geographical origin of this sapphire is limited; as the only known gem quality of important size blue sapphire deposits before 1669 were where is today Myanmar, Sri Lanka and Thailand-Cambodia (see Chap. 2). UV-Vis spectra are similar to those acquired on blue sapphires from Myanmar and Sri Lanka as those from Thailand-Cambodia are basalt related and as a consequence present strong absorption around 800 nm. Growth zoning as well as the long and thin rutile needles lead the authors to draw the conclusion that the stone most likely originates from Sri Lanka.

5.4 Other Questions

Gemmology, or methods applied on gems, can also help to answer other questions. For example, two historic blue diamonds, the Hope (45.52 cts, antique cushion cut) and the Wittelsbach-Graff (31.06 cts. cushion modified brilliant cut) were examined in order to check if they were cut from the same piece of rough (Gaillou et al. 2010). Blue diamonds are rarer than colourless diamonds. These two blue large diamonds were once part of the crown jewels of European monarchies. The Hope diamond once belonged to the French Royal Family and the Wittelsbach-Graff blue was owned by the Bavarian Royal Family, the House of Wittelsbach. Previously, some publications presented the possibility that the Hope diamond was cut from an approximately 115 ct stone (the Tavernier Blue purchased in India possibly at the Kollur mine, Golconda area) that Jean-Baptiste Tavernier sold to Louis XIV of France in 1668 (Farges et al. 2009). It is also believed that the Wittelsbach-Graff blue comes from the Kollur mine, Golconda, and that it arrived in Vienna in 1666 as part of a dowry for a marriage into the House of Habsburg and in 1722 passed to the House of Wittelsbach, again as a part of a dowry (Dröschel et al. 2008). Infrared spectra of the two diamonds showed that they are both Type IIb; *i.e.*, their boron exceeds nitrogen concentration, if any. Estimation of the uncompensated boron concentration from the infrared absorption coefficient yielded boron concentration of 0.36 ± 0.06 ppm (atomic) for the Hope and 0.19 ± 0.03 ppm for the Wittelsbach-Graff. Both diamonds exhibited intense orange-red phosphorescence, visible to the unaided eye in a dark room for approximately 1 min. However, the mosaic patterns observed under the Diamond View (ultra short-wave UV excitation) were different; coarser in the Wittelsbach-Graff (mainly features >200 microns) than in the Hope (mainly features <100 microns). The results indicated that the two diamonds did not originate from the same crystal, although they likely experienced similar geologic histories.

Other recent methods included radiocarbon (^{14}C) dating which is applied to calculate the age of pearls (Krzemnicki and Hajdas 2013; Zhou et al. 2017). This technique requires a significant amount of material (> 5 mg) to be extracted and analyzed and may have limitations with pearls of unknown origin as corrections linked to the local reservoir need to be done. Freshwater pearls may pose additional challenges. However, this method can give fairly good results for saltwater pearls of known origin formed before 1955. A set of pearls was reportedly harvested from the Caribbean islands off the coast of Venezuela during the early sixteenth century, most likely from the waters off Cubagua Island. Carbon dating on different samples using three different instruments gave calibrated ages (assuming they were fished from the waters off Cubagua island) to be between late fifteenth century and early seventeenth century. So, some of the pearls are possibly fished before or at the start of the Columbian Era (Zhou et al. 2017).

References

Barone G, Bersani D, Jehlička J, Lottici PP, Mazzoleni P, Raneri S, Vandenabeele P, Di Giacomo C, Larinà G (2015) Nondestructive investigation on the 17-18th centuries Sicilian jewelry collection at the Messina regional museum using mobile Raman equipment. J Raman Spectrosc 46:989–995

Bersani D, Andó S, Vignola P, Moltifiori G, Marino IG, Lottici PP, Diella V (2009) Micro-Raman spectroscopy as a routine tool for garnet analysis. Spectrochim Acta A 73:484–491

Calligaro T, Mossmann A, Poirot JP, Querre G (1998) Provenance study of rubies from a Parthian statuette by PIXE analysis. Nucl Inst Methods Phys Res B 136-138:846–850

Dröschel R, Evers J, Ottomeyer H (2008) The Wittelsbach Blue. Gems Gemol 44:348–363

Farges F, Sucher S, Horovitz H, Fourcault JM (2009) The French Blue and the Hope: new data from the discovery of a historical lead cast. Gems Gemol 45:4–19

Farges F, Panczer G, Benbalagh N, Riondet G (2015) The Grand Sapphire of Louis XIV and the Ruspoli sapphire. Gems Gemol 46:80–88

Gaillou E, Wang W, Post JE, King JM, Butler JE, Collins AT, Moses TM (2010) The Wittelsbach-Graff and Hope diamonds: not cut from the same rough. Gems Gemol 46:80–88

Giuliani G, France-Lanord C, Coget P, Schwarz D, Cheilletz A, Branquet Y, Giard D, Martin-Izard A, Alexandrov P, Piat DH (1998) Oxygen isotope systematics of emerald: relevance for its origin and geological significance. Mineral Deposita 33:513–519

Giuliani G, Chaussidon M, Schubnel HJ, Piat D, Rollion-Bard C, France-Lanord C, Giard D, de Narvaez D, Rondeau B (2000) Oxygen isotopes and emerald trade routes since antiquity. Science 287:631–633

Giuliani G, Chaussidon M, France-Lanord C, Savay Guerraz H, Chiappero PJ, Schubnel HJ, Gavrilenko E, Schwarz D (2001) L'exploitation des mines d'émeraude d'Autriche et de la Haute Egypte à l'époque Gallo-Romaine: mythe ou réalité? Revue de Gemmologie AFG 143:20–24

Groat L, Giuliani G, Marshall D, Turner D (2014) Emeralds. In: Geology of Gem deposits, vol 44. Mineralogical Association of Canada, Québec, pp 135–174

Hänni H, Schubiger B, Kiefert L, Häberli S (1998) Raman investigation on two historical objects from Basel cathedral: the reliquary cross and Dorothy Monstrance. Gems Gemol 34:102–125

Karampelas S, Wörle M, Hunger K, Hanspeter L, Bersani D, Gübelin S (2010) Study of gems of a ciborium from Einsiedeln Abbey. Gems Gemol 46:291–295

Karampelas S, Wörle M, Hunger K, Hanspeter L (2012) Micro-Raman spectroscopy on two chalices from the Benedictine Abbey of Einsiedeln: identification of gemstones. J Raman Spectrosc 43:1833–1838

Krzemnicki MS, Hajdas I (2013) Age determination of pearls: a new approach for pearl testing and identification. Radiocarbon 55:1801–1809

Rondeau B (2003) Matériaux gemmes de référence du Museum National D'Histoire Naturelle: exemples de valorisation scientifique d'une collection de minéralogie et gemmologie. PhD thesis, University of Nantes, Nantes, France

Saeseaw S, Pardieu V, Sangsawong S (2014) Three-phase inclusions in emerald and their impact on origin determination. Gems Gemol 50:114–132

Schwarz D, Pardieu V (2009) Emeralds from the Silk Road countries. A comparison with emeralds from Colombia. InColor 12(Fall/Winter):38–43

Smith CP, Darenius EQ (2009) Inside emeralds. Rapaport Diamond Rep 32:139–149

Zhou C, Hodgins G, Lange T, Saruwatari K, Sturman N, Kiefert L, Schollenbruch K (2017) Saltwater pearls from the pre- to early Columbian era: a gemological and radiocarbon dating study. Gems Gemol 53:286–295

Correction to: Gems and Gemmology

Correction to:
S. Karampelas et al., *Gems and Gemmology,*
Short Introductions to Cultural Heritage Science,
https://doi.org/10.1007/978-3-030-35449-7

The original version of the book was revised to replace a number of figures as follows:

- new figures with new figure captions appear in Chapters 2, 3 and 4: 2.3, 2.6, 2.7, 2.8, 2.9, 2.12, 2.13, 2.15, 2.16, 2.18, 2.20, 2.25, 2.27, 2.28, 2.33, 3.1, 3.2, 3.3, 3.4, 3.5, 3.6, 3.8, 3.9, 3.12, 3.14, 3.15, 3.16, 4.5, 4.10.

- new figures but with unchanged captions appear in Chapters 2, 3, 4 and 5: 2.1a, 2.1b, 2.2, 2.4, 2.5, 2.10a, 2.10b, 2.11, 2.14, 2.17, 2.19, 2.21, 2.22, 2.23, 2.24, 2.26, 2.29, 2.30, 2.31, 2.32, 2.34, 3.7a, 3.7b, 3.10, 3.11, 3.13, 4.1, 4.2, 4.3a-c, 4.4a, 4.4b, 4.6, 4.7a, 4.7b, 4.8a, 4.8b, 4.9, 4.11, 5.1, 5.2, 5.3, 5.4a, 5.4b, 5.5, 5.6, 5.7, 5.8.

The biography of Stefanos Karampelas has been updated, as well as his affiliation. Further, the Acknowledgements section now includes the names of those who took the photos for the replacement figures.

The updated version of the book can be found at
https://doi.org/10.1007/978-3-030-35449-7
https://doi.org/10.1007/978-3-030-35449-7_2
https://doi.org/10.1007/978-3-030-35449-7_3
https://doi.org/10.1007/978-3-030-35449-7_4
https://doi.org/10.1007/978-3-030-35449-7_5

Glossary

Amorphous (gem) has a specific chemical composition but without a definite crystal structure. For example, opal's chemical formula is $SiO_2(nH_2O)$ and it is mostly not crystalized (amorphous).

Asterism (gem) is a star-like phenomenon caused by light interacting with needle/tube-like inclusions (example rutile like mineral inclusions or empty tubes). Gems are cut as cabochons to reveal this effect. The number of rays (4-rays, 6-rays, 12-rays) depends on the orientation of the inclusions and the cut. **Chatoyancy** (or cat's eye) is when only one line is shown.

Autoclave is an apparatus with a chamber in which the pressure and temperature can be altered. It is also used to synthetize gems.

Carat (gem) is a unit of weight equivalent to 0.2 grams.

Chromophore is the element responsible for colour (colouring element). A gem is **allochromatic** when the chromophore is an impurity (*i.e.*, not essential of gem's chemical composition); so, this gem is colourless in its pure state (e.g., diamond, beryl, corundum, quartz, spinel). **Idiochromatic** is a gem which is coloured due to a chromophore (chemical element) essential for its chemical composition (e.g., peridot, some garnets).

Cloisonné is a technique for decorating objects. It was used to decorate jewellery pieces with gems used as inlays with gold or silver around them used as compartments (cloisons in French). Garnet cloisonné was used in jewellery before the eighth century C.E.

Cryptocrystalline/microcrystalline/massive (gem) consists of aggregates not visible to the unaided eye. For example, chalcedony is the cryptocrystalline/cryptocrystalline/massive variety of quartz (SiO_2; trigonal crystal system).

Crystal structure is the ordered arrangement of the chemical elements (atoms, ions or molecules) in the gem in a particular (unique) way (lattice) which is periodically repeated in three dimensions. For every crystalized gemstone (or mineral in general) there are only 7 ways of arrangements/crystal systems called crystal systems. The seven crystal systems are: cubic, tetragonal, trigonal, hexagonal, orthorhombic, monoclinic and triclinic. Crystal systems have crystallographic axis, specific for each crystal system; for example, in cubic system (diamond) there are 3 axes all of equal size and in hexagonal (beryl-emeralds-) there are four axes with two axes' sizes with the optic axis or c-axis the longest and the other three shortest with equal size. Crystal systems are related to the optic character of the gem; *i.e.*, how the light travels along the crystallographic axis. Crystalized gems are either isotropic (with all the axes equal and the light travelling exactly the same way, so the gemstones crystalized to the cubic system; e.g., diamond, garnet, spinel) or anisotropic (gemstones crystallized to the other 6 crystallographic systems). Anisotropic gemstones are divided to uniaxial (two crystallographic axes lengths -one is the optic axis-) are those crystalized to tetragonal, trigonal and hexagonal system; so the light travelling in two different ways along the gemstone; e.g., zircon, corundum -ruby, sapphire-, beryl -emerald, aquamarine-) and biaxial (three crystallographic axis lengths and two optic axes) are those crystalized to orthorhombic, monoclinic and triclinic system; so the light travelling in three different ways along the gemstone; e.g., olivine -peridot-, chrysoberyl -alexandrite-, zoisite -tanzanite-, topaz). The anisotropic stones are also double refractive, *i.e.*, the light will break in different rays and will travel at different speeds with the gemstone.

Doublet (gem) is composed of two sections glued together and used as imitations. **Triplet** is composed by three sections glued together (commonly the intermediate layer is coloured) together and used as imitations.

Gems are the materials used for adornment or decoration which are relatively rare, hard and tough. Gems can be (single) crystals (minerals) of one species or of solid series, some are amorphous, other are rocks and some are composed partly or entirely of organic materials.

Hardness (gem) is the ability to resist abrasion by other materials. Mohs hardness is a scale used to measure hardness relatively and is composed of ten minerals with hardness from 1 to 10 as follows: Talc (1), Gypsum (2), Calcite (3), Fluorite (4), Apatite (5), Feldspar -Orthoclase- (6), Quartz (7), Topaz (8), Corundum (9) and Diamond (10). Fingernails are considered to have hardness of about 2.5 (can scratch gems -or materials- with hardness of talc and below and it can be scratched by gems -or materials- with the hardness of calcite and above), the blade of a knife about 5.5 and glass around 6.5. Hardness is different to **toughness** which roughly reflects the mechanical resistance.

HPHT (gem) refers to High Pressure High Temperature. It is a treatment of diamonds when the conditions are approaching the conditions of diamond's formation (HPHT treated) as well as a method to synthetize diamonds (HPHT synthetic diamonds).

Imitations (gem) is a material that looks similar to a gem but has different chemical composition and crystal structure.

Inclusions (gem) crystals, liquid or gas filled cavities enclosed into a gem.

Irradiated is exposed to radiation. Gems are sometimes artificially irradiated in order to enhance their appearance. Irradiation is done with electrons, gamma rays or cobalt-60.

Glossary

Isotope is a variant of a chemical element which differs in neutron number. Isotope analysis is the identification of the ratio between different variants of the same chemical element (e.g., hydrogen, carbon, oxygen, nitrogen, sulfur).

Luminescence is the emission of the light due to radiation (e.g., UV light and X-rays). It includes, between others, fluorescence and phosphorescence. The reaction of a gem under ultraviolet (UV) lamp radiation is called fluorescence (under UV light) and its colour and intensity are used to describe it. In case the reaction continues when the source/excitation (UV light) is off is called phosphorescence (under UV light) and its colour, intensity and duration are used to describe it. Short-wave UV (SWUV) lamps emitting at 254 nm and long-wave UV (LWUV) lamps emitting at 365 nm (from 3 to 6 Watt) are used in gemmology.

Lustre (gem) is a surface characteristic which depends on the gem's R.I., its polish and surface condition. In gemmology the terms adamantine (e.g., diamond), sub-adamantine (e.g., zircon), vitreous (e.g., emerald), resinous (e.g., amber), waxy (e.g., turquoise), silky (e.g., malachite), pearly (e.g., pearl), and metallic (e.g., hematite) are used to describe lustre.

Organic (gem) is the gem partially (and the rest mineral) or entirely composed of organic materials. All organic gems are not forcibly biogenic; e.g., jet is organic without being biogenic.

Organogenic/biogenic (gem) is an organic gem formed from or with the help of a living organism (e.g., natural pearl from a mollusc).

Pleochroism (gem) is the appearance of more than one colour in the same gem when rotated. It is based on the crystallographic orientation of the gem when observed. Isotropic gems do not present pleochroism, it is a characteristic of the anisotropic gems. In uniaxial gems two colours are appearing (dichroism) and in biaxial three (trichroism). Pleochroism is different from colour change (or alexandrite effect) which is observed when colour differs while viewed under different lighting (under daylight (5500 K) and under incandescent light (3600 K)).

Polarized light is travelling of light only in one direction. Light's polarization is done using a polarizer which allows light to pass through one direction.

Polymorphs (gems) are of the same chemical composition but with different crystal structure. For example, diamond and graphite are polymorphs, the first is crystalized in a cubic crystal system and the second in a hexagonal crystal system.

Primary mine means that gem is found in situ (the host rock in which it has formed). When a gem is found in a **secondary mine** this means that it is found out of the host rock where it was formed, e.g., in the river after being removed naturally (by water) from its initial location.

Radiocarbon dating is a method used for age determination of organic materials by using the properties of radiocarbon (^{14}C, radioactive isotope of carbon); a.k.a. carbon dating or carbon-14 dating.

Radioactivity occurs when particles are emitted from a nuclei due to nuclear instability.

Reflection happens when light falls on a surface and partially or fully comes back (reflected).

Refraction is the bending of light as it is passing from one mean to another with different refractive index (e.g., air to water). **Diffraction** refers to the phenomenon occurring when the light passes through a slit (spread out). **Interference** refers to the phenomenon occurring when two waves are superimposed. **Dispersion** is the separation of the light when it passes through a prism (or another medium) and split in the different colours of the electromagnetic spectrum. **Scattering** (light) is occurring when light is deviating due to particles when passes through.

Refractive index (gems): It signifies the degree the light (measured with a yellow light −589.6 nm-) is bent when it enters a stone. This is measured with a refractometer. For example, spinel has a refractive index (R.I.) of around 1.74 and diamonds 2.42, that means they bend light 1.74 and 2.42 times more than air respectively. Isotropic gems have one refractive index and anisotropic (double refractive) two (uniaxial) or three (biaxial). Both uniaxial and biaxial can be positive and negative. Uniaxial positive is when the moving refractive index is greater than the fixed refractive index value and uniaxial negative when the moving refractive index is lower than the fixed refractive index value. Biaxial positive is when the intermediate refractive index is closer to the higher and biaxial negative when the intermediate refractive index is closer to the lower. The difference between the minimum and maximum R.I. is the birefringence. Spot reading is a method to estimate the R.I. if a gem is cut as cabochon or is opaque, which differs from that used for transparent gems. Dispersion of gems is the difference of R.I. measured with red light (686.7 nm) and violet (430.8 nm); e.g. diamond has high dispersion (0.044). Fire of a gem is linked to the dispersion; it is more pronounced when dispersion is higher.

Rock (gem) is an assemblage, solid aggregate, of one or more minerals. Jadeite "jade" gem is a rock consisting principally of one mineral (jadeite) and lapis lazuli gem is a rock consisting of various minerals (e.g., mainly lazurite but also pyrite, calcite and sodalite).

Rocks (general) are divided to igneous, metamorphic and sedimentary. Igneous (or magmatic) rock are formed through the cooling of magma or lava. Sedimentary rocks are formed by erosion of other, mineral (or rock) depositions and cementation of them along with organic particles on the floor of oceans or other bodies of water at the earth surface. Metamorphic rocks are formed from the transformation of existing rock types (e.g., igneous, sedimentary or existing metamorphic rock) due to variations of heat and pressure.

Rough (gem) is a gem as found in nature before faceting (cutting) or carving.

Single crystal (gem) is relatively homogenous with a specific chemical composition (only varying at impurities level) arranged in an orderly manner (specific crystal structure; i.e., crystalized in one of the crystal systems). For instance, diamond is made of carbon and is crystalized in the cubic crystal system.

Solid solution series (gem) has a specific crystal structure but with slightly variable chemical composition. For example, peridot (green variety of olivine) in crystalized in orthorhombic crystal system and has the following chemical formula $(Mg,Fe)_2SiO_4$, with forsterite (Mg_2SiO_4) and fayalite (Fe_2SiO_4) the endmembers.

In nature a mix of the end members is found (e.g., 90% Forsterite and 10% Fayalite); virtually never a pure end member can be naturally found.

Softening (gem) is a treatment historically used to soften gems (mainly diamonds) using different means in order to become easier to cut/engrave.

Spectroscopy (gem) is a method using a spectrometer of measuring on how light of particular wavelength (or wavenumber) can interact with the gem. Spectral resolution measures the ability to resolve features in a spectrum.

Synthetic (gem) is laboratory grown gem with essentially the same chemical composition (differences between natural and synthetic gems are at impurities level) and crystal structure.

Transition metals are the chemical elements of groups 4–11 at the periodic table. The most important in gemmology are these in the first row which are commonly present as gems' chromophores (Ti, V, Cr, Mn, Fe, Co, Ni and Cu).

Treatments (gem) are all the processes used to modify gem's appearance in order to make them more attractive.

Vernier calliper is a measuring tool used sometimes for gems.

Index

A
Agate, 5, 11, 34, 68, 74, 96
Amber, 1, 5, 26–29, 44–45, 47, 50, 54, 57, 76, 77, 79, 81, 82
Amethyst, 5, 9, 11, 33, 75, 81, 86, 96
Ametrine, 33

B
Beryl, 9, 21–23, 50, 75, 80, 86, 100

C
Cameo, 6, 7, 9, 13, 24, 26, 33, 34
Carnelian, 9, 34, 68, 96
Chalcedony, 5, 9, 33, 34, 44, 45, 74, 96
Chatoyancy/asterism, 31, 33
Chemistry, 1, 29, 59, 63, 74, 83, 101
Chrysoberyl, 29–32, 85
Chrysoprase, 33
Citrine, 33, 47, 75, 86, 95
Cloisonné, 24
Coating, 42, 63, 68, 70–75, 84
Colour change, 13, 31, 85
Coral, 26–29, 44, 50, 52, 54, 63, 70, 75, 81, 88
Corundum, 9, 13–16, 54, 57, 62, 63, 68–70, 75–77, 79, 80, 83, 85, 87, 94, 95
Cultured pearls, 12, 42, 52, 56, 60, 74, 76, 80, 84
Cut, 1, 3, 5, 7–9, 12, 14, 16, 18, 21, 25, 39, 40, 63, 78, 84, 92, 96, 102

D
Diamonds, 2, 3, 5, 7–9, 18, 20–21, 25, 39, 43, 45, 48–50, 52, 56, 57, 68, 69, 72, 79–81, 83, 87, 88, 99, 102

Dyeing, 3, 67, 68, 70, 72–76, 81, 83

E
Emeralds, 3, 5, 9, 11, 21–23, 31, 39, 42, 49–50, 52–54, 57, 59, 61–63, 68, 70, 72, 74, 75, 80–82, 85–87, 100, 101

F
Faceting, 5, 6, 9, 13, 23, 32, 43, 44, 70, 71, 79, 84, 96
Foiling, 3, 70, 72–75

G
Garnets, 9, 13, 24–25, 43, 45, 49, 54, 55, 63, 76, 83, 85, 92, 94–96, 100
Gem materials, 1, 3, 17, 20, 26, 34, 40, 42, 45, 46, 54, 67, 73
Gemmology, 1, 19, 39–49, 54, 56–58, 61, 102

H
Heat treatment, 68, 69, 75–80

I
Imitations, 1, 2, 49, 54, 58, 67–88, 92, 95–100
Intaglio, 6, 7, 33
Irradiation, 69, 80
Ivory, 26–29, 42, 44, 60, 63, 81, 83

J
Jade, 5, 11, 19–20, 52, 73, 75, 81

© Springer Nature Switzerland AG 2020
S. Karampelas et al., *Gems and Gemmology*, Short Introductions to Cultural Heritage Science, https://doi.org/10.1007/978-3-030-35449-7

Jadeite, 19, 20, 50, 52, 63, 70, 73, 75, 83
Jasper, 34

L

Lapis lazuli, 5, 11, 13, 16–19, 50, 63, 75, 81, 88

M

Microscopes, 2, 3, 28, 33, 40–42, 52–53, 56–58, 61–63, 70, 72, 73, 75, 76, 79–86, 91, 94–96
Milky quartz, 33

N

Natural pearls, 3, 10–12, 56, 76, 84
Nephrite, 5, 19, 20

O

Odontolite, 83
Oiling, 68, 70, 72–75, 97
Onyx, 6, 34, 74, 81, 96
Opal, 5, 35, 45, 50, 54, 74, 82, 83, 88

P

Pearls, 1, 3, 5, 10–13, 25, 26, 39, 42, 44, 50, 52, 54, 56, 59, 60, 63, 68, 69, 72, 74, 76, 80, 81, 84, 91, 92, 94, 102
Peridot, 9, 29–31, 49, 50, 95–97
Play of colour, 35, 50
Prasiolite, 33

R

Refractive index (R.I.), 42–47, 73, 77, 81, 83
Rock crystal, 9, 33

Rose quartz, 33
Ruby, 5, 13–16, 24, 30, 40, 48, 49, 53, 68, 70, 73, 75–78, 81, 82, 84, 88, 92, 95, 97, 101

S

Sapphire, 3, 5, 13–16, 39, 40, 45, 48, 50, 52, 67–70, 75, 77–82, 85, 86, 88, 94, 96, 97, 100, 101
Sard, 34, 68
Sardonyx, 33
Silica gems, 6, 9, 33–35
Smoky quartz, 33, 86, 96
Specific gravity (S.G.), 43–47, 82, 83
Spectroscopy, 3, 39, 48–49, 50–57, 61–63, 73–76, 79, 86, 88, 96, 100, 101
Spinel, 13, 24, 29–31, 43, 45, 49, 50, 53, 56, 61, 76, 83, 85, 86, 96, 97
Synthetic, 1–3, 39, 42, 45, 48–50, 52, 54, 57, 59, 63, 67–88, 92

T

Topaz, 9, 29–31, 47, 50, 68, 69, 71, 72, 75, 76, 80, 95–97
Tourmaline, 5, 29–31, 40, 44, 49, 75, 79, 81, 88
Treatments, 1–3, 40, 46, 48, 52, 56, 61, 67–88, 92, 97
Turquoise, 9, 16–18, 22, 44, 54, 70, 73, 75, 81, 83, 88, 97

X

X-ray imaging, 3, 60–61

Z

Zircon, 25, 29–32, 47, 49, 75, 77, 83, 92, 94

The manufacturer's authorised representative in the EU is Springer Nature Customer Service Centre GmbH, Europaplatz 3, 69115 Heidelberg, Germany. If you have any concerns regarding our products, please contact ProductSafety@springernature.com

Printed and bound by CPI Group (UK) Ltd, Croydon, CR0 4YY

25/03/2026

02078177-0011